PYGMALION

畢 馬 龍 P Y G M A L I O N

令人心醉惑溺的女性人物模型塗裝技法

如此美麗，讓人不禁陷入戀情的魅惑女性形象 · 田川 弘塗裝作品 A to Z

Obsessive finish of Girl's figure.
The A to Z of Hiroshi Tagawa's
fascinating female images.

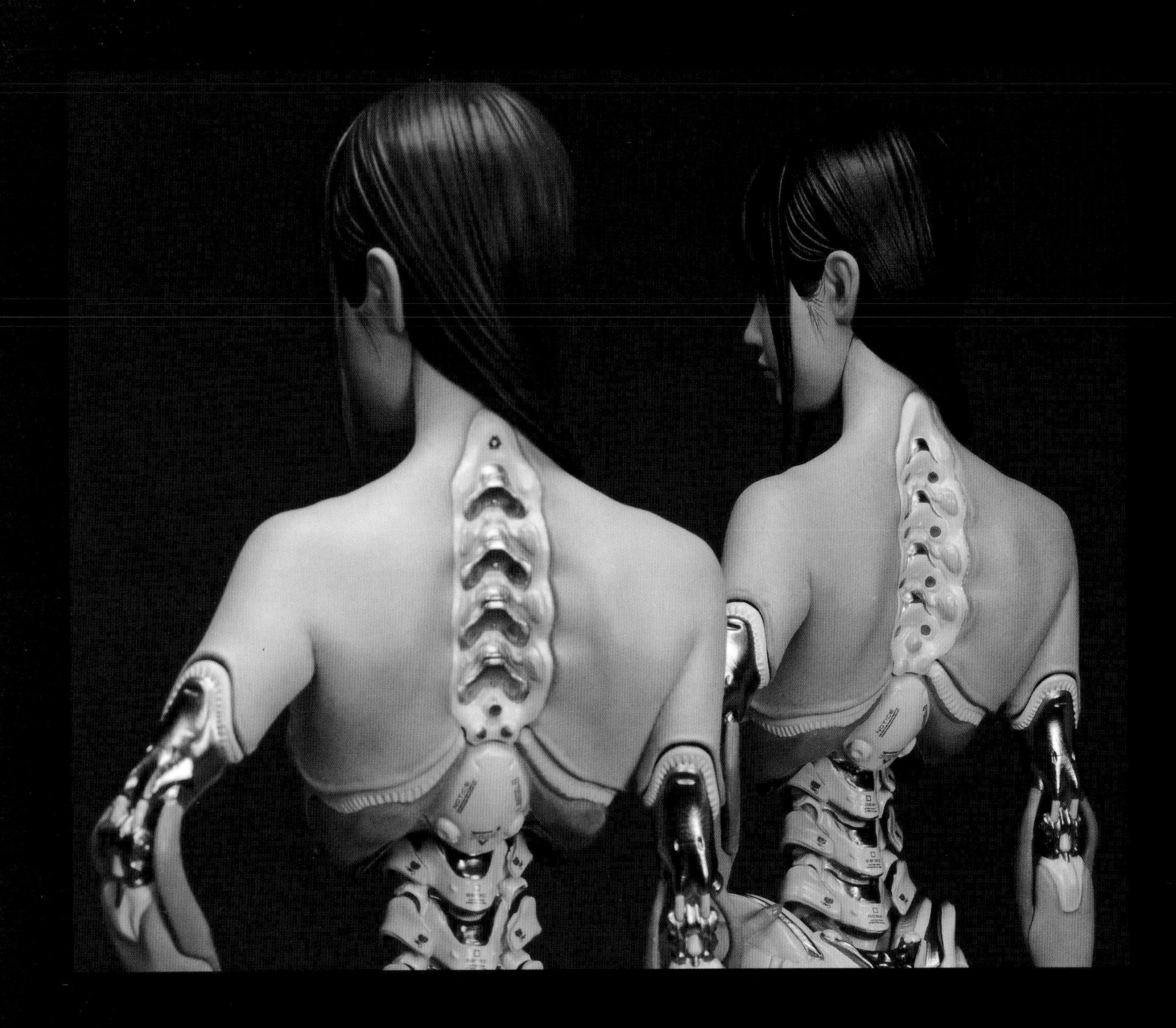

Android EL01

2019 年 1/6 比例 3D 列印輸出套件 原型 K

2019 1/6 Scale 3Dprinter output product Sculpted by K

此為第 38 期「Wonder Showcase」的推廣作品，由現職產品設計者所製作的原創作品。套件的名稱雖然叫做「Android」（人型機器人），但田川老師第一眼見到即心有所感，決定要將她視作「Cyborg」（化人）來賦予她靈魂。製作時沒有太多改造的部位，只有植上睫毛而已。機械骨骼的部分貼上細微的英文字母水貼裝飾，呈現出「剛製造出來的工業產品」的印象，也形成引人注目之處。此外，在身體的底層塗裝有描繪出血管。

This is the 38 season presentation of Wonder Show Case, an original work by an active product designer. The name of the kit is "Android" but Tagawa felt the spirit at first sight and wanted to give soul as "cyborg". There was no particular modification to the product, and only eyelashes were transplanted. The frame of the machine is decorated with fine letter decals to evoke an image of "nascent industrial product" as an eye catch. Also, blood vessels are drawn all over the body as a foundation coating.

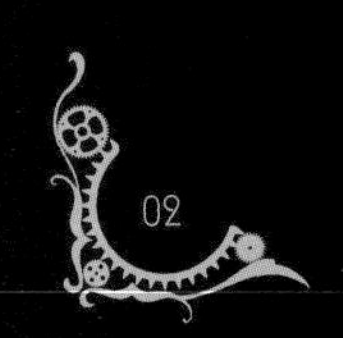

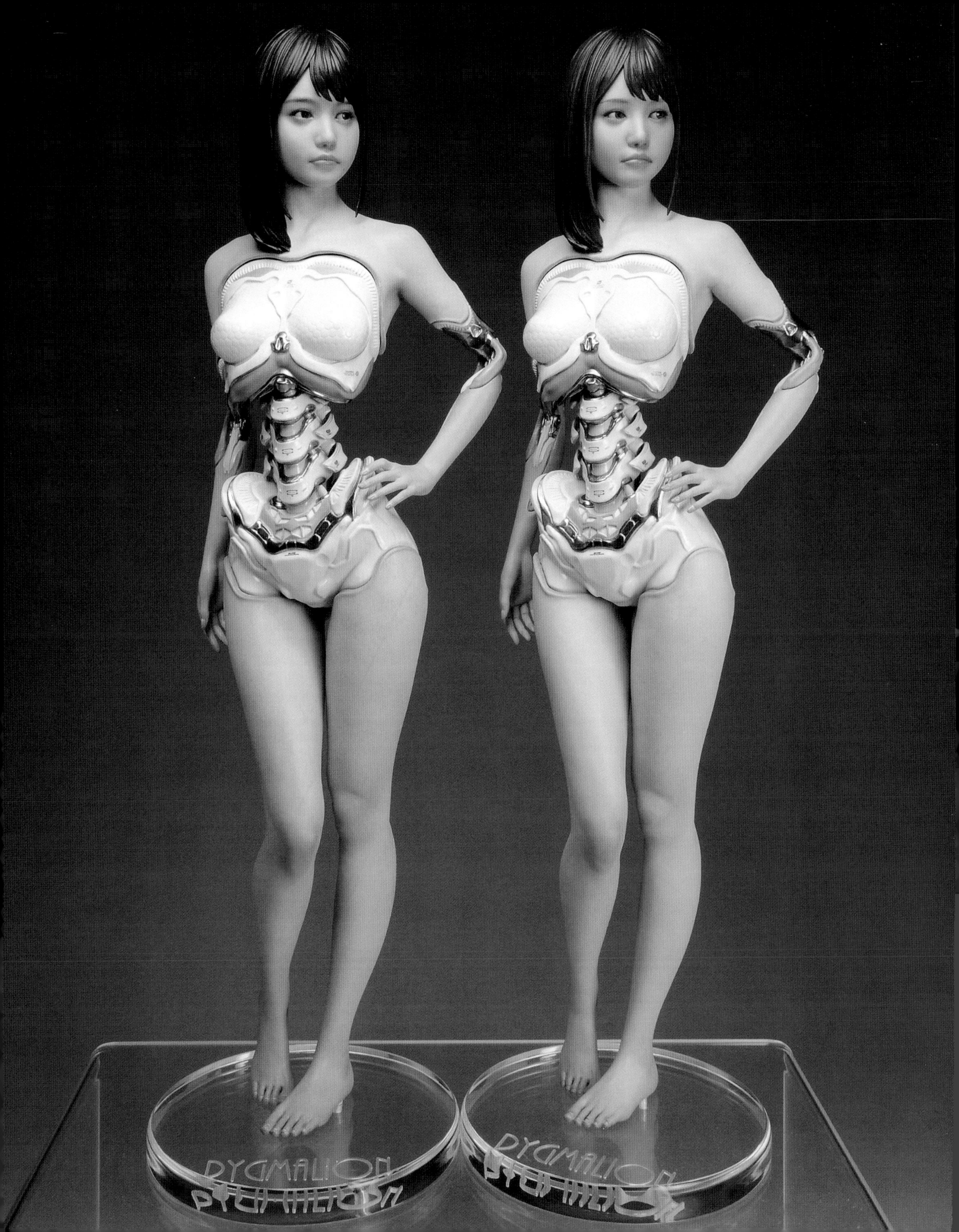
PYGMALION
PYGMALION

Which?

which-02

2020 年 1/12 比例 Which？
樹脂鑄型套件 原型 林 浩己

2020 1/12 Scale which02 Resin cast Kit
Sculpted by Hiroki Hayashi

氣質高雅，足以登上女性時尚雜誌封面的 SM 女王。作品甚至透露出可以讓人感受到遭她鞭打時的衝擊感。上衣改短，增加過膝皮靴。再以蝕刻片追加帽章。

She is a stylish S&M girl who might appears on the cover of a fashion magazine. It was a shock like being whipped by her. The jacket was shortened and the boots was converted into knee-high boots. Added a hat pit with photo-etched parts.

INDEX

※與作品名稱併記的年份為作者完成作品當時的西元年。

Black Rock City

2019 年　無比例
樹脂鑄型套件 原型 大畠雅人

2019 Non-Scale Resin cast Kit Sculpted by Masato Ohata

在她的視線中，熊熊燃燒著絕不放棄的生命之焰。RPG7 是由中國製 1/6 來福槍集合套件中取用。

The gaze sparkle with fire of life. The RPG7 is a kit from the Chinese 1/6 rifle collection.

田川 弘的人物模型製作心得 1

Tagawa's knowledge of figure making 1

無論造形物本身是否已經具備了壓倒性的優秀表現力，只要田川 弘心裏還沒有設定好這個立體物件所處的情節背景，就不會實際去進行塗裝作業。這個情節背景，是用來找出她們所處的世界與我們這個世界的交接點，讓我們能夠去接觸她們，並且逐漸融入她們的世界。理解並融入情節背景，是開始製作人物模型的切入點，同時也是激發塗裝人物模型意願的原動力。唯有明確地定義出作品所處的背景環境以及情景狀態，才能夠催生具有生命力的作品。

Even if you come across a molded objects with a superior and overwhelming expressiveness, if you can't create a story within yourself from it, you can't move on to actual painting. The creation of a story is nothing but finding a connection with the world in which they live, touching it, and gradually getting into it.The story is not only a starting point for making a figure, but also a driving force for painting a figure at the same time. A powerful work can be created when there is a clear story that forms the backbone of the work.

田川 弘的人物模型製作心得 2

Tagawa's knowledge of figure making 2

我們要追求的不是製作出一件精美而且完成度高的模型。即使外觀看起來和真人的外形一模一樣，但擺放在我們面前的只是一個叫做樹脂鑄型套件的物體。那麼要如何才能讓她們的血液流動起來？怎麼樣才能讓她們的肌膚、秀髮得到柔軟感覺以及活力呢？要怎麼做才能讓她們散發出耀眼的光芒？想要找出這個答案，田川 弘能夠倚靠的只有自身所擁有的技巧，也就是「描繪」這個手段而已。讓熱血在零件裡流動，讓人能夠感受到如真人肌膚般的溫暖，並在她的眼中注入靈魂意志。對田川 弘來說，描繪這個動作的意義，就是在於描繪生命。塗裝套件的這個行為，是為了要表現出這個套件所擁有的生命力，田川 弘便是懷抱著如此的強烈熱情，不斷地描繪下去。

It's not about making good things, or models with high perfection. There's an object called a resin kit. How can their skin and hair achieve softness? How can they shine? The only technique he has is "drawing". The blood runs through the parts, making you feel the warmth of human skin and putting your will into the eyes. For Tagawa, drawing means drawing life. Draw strongly that the act of applying a kit is to express the vitality of the kit.

田川 弘的人物模型製作心得 3

Tagawa's knowledge of figure making 3

對於自己正在塗裝的女性，田川 弘在作業時總是懷抱著「戀愛」的感覺。與自己所感受到的，她們所身處的情節背景形成共鳴，有時甚至因為感動而潸然流淚。擺放在面前的樹脂鑄型套件這個「冷冰冰」的物體，經過肌膚、頭髮、嘴唇、指甲及眼神的描繪之後，搖身一變成為了一位「活生生」的少女。「將靈魂傾注在她身上」這絕非虛無飄渺的空話，而是真實滿懷熱情能量面對作品的態度。所以，有非常多人在看到田川 弘完成的作品之後，都不禁說出「愛上她了」的感想。由田川 弘所賦予的靈魂形態，使得其他人也能夠獲得共鳴，而這正是將靈魂傾注其中的意義所在。

Feel the "love" while painting the girls. He sympathizes with their stories and sometimes burst into tears. In the process of drawing the skin, hair, cheeks, lips, nails, and eyes on an object called resin figure, it transforms into a girl who "exist" there. It is not just about putting your soul into the work, it is about challenging the work with real energy. For this reason, when people look at the completed Tagawa's works,"Fell in love". Other people feel sympathy for their soul given by Tagawa, the soul he gave to make the figure something meaningful.

在此想要以客觀的角度來為各位介紹本書的作者田川 弘。
不怕各位誤解，田川 弘他熱愛人偶（人物模型）。這個行為本身屬於個人的嗜好，並非什麼特別奇怪的事情。比方說有人會把自己的愛車當作「夥伴」，小心翼翼地呵護，甚至有時還會對愛車說說話。田川 弘對於人偶的熱愛，與這樣的行為並沒有什麼不同。
田川 弘從小生長的環境就圍繞在母親因興趣製作的法國娃娃當中，他自己也喜歡製作組合模型，然後接觸到繪畫，於是便以此為人生志向。高中與大學都進入美術科系，專攻西洋繪畫。在當年便以還算少見的插畫筆觸，描繪出以年輕女性為描繪主題的真人尺寸寫實畫作。當時他的口頭禪是「我不是在描繪一個真人的女性。而是在畫布裡面製作一個人偶（人物模型）」。田川 弘 28 歲時首次參加公募展，旋即獲得首獎。然而之後遭遇到極大的瓶頸，10 年後甚至放下了畫家的畫筆。在這樣的狀態下，田川 弘偶然邂逅了原型師．林 浩己的「擬真女性裸體人物模型」。這個人偶，不就是自己一直在追求的立體畫布嗎？於是田川 弘便開始大量收購女性人物模型，並瘋狂地為人偶進行塗裝。20 年前田川 弘開設的個人網頁的作品藝廊，也在很短的時間內由繪畫作品轉為以人物模型為中心的展示內容。
田川 弘在塗裝人物模型時，首先會由自己為其設定一個背景情節、一場故事。如果無法將故事設定完成的話，就不會進入塗裝的作業。一旦進入塗裝作業後，甚至會因為內心豐富的情感向外滿溢而出，導致常常一邊流著淚水，一邊進行塗裝。將情感帶入到這個程度，已經可以說是進入了無限接近戀愛情感的境界了。希臘神話中，賽普勒斯島的國王畢馬龍愛上了自己親手製作的象牙雕像，便向愛神阿芙蘿黛蒂祈求使雕像變成真人，並與她結為夫婦。田川 弘對於自己製作出來的人物模型的熱情，也許無法傳遞至天上的阿芙蘿黛蒂，但確實擁有能夠讓地上的許多人為之傾心動搖，讓所有看到作品的人都在心中萌生「愛意」的力量。

Without being afraid of misunderstandings, they are one of the ideas, not particularly peculiar. For example, it is no different from treating your car "buddy", treating it with care, and sometimes talking to it. Since his childhood, Hiroshi Tagawa enjoyed making plastic models surrounded by French dolls his mother made as a hobby. He entered high school, university and art course and majored in Western painting. He drew a realistic, life-sized young woman with a touch of illustration, which was rare in those days. His pet phrase at that time was "It's not about women as human beings. I'm making a figure in a canvas.". At the age of 28, he participated in an exhibition sponsored by the public for the first time, and he suddenly won the grand prize, but after that, he ran into a big obstacle, and 10 years later, he gave up painting as a professional. It was during this time that he encountered the prototypical teacher, Hiroki Hayashi's 'Real Female Nude Figure'. He thought that this doll was really a three-dimensional canvas, and after that, he devoured her figurines and painted them on. It did not take long for the gallery of the website, which was opened 20 years ago, to focus on figures from paintings.
Tagawa says that when he paints figures, he first thinks of their backbone stories. If that is not decided, the painting process cannot proceed. Once you start painting, sometimes you get emotional and continue painting with tears running down your face. That level of empathy is infinitely close to romance. According to Greek mythology, Pygmalion, the king of Cyprus, fell in love with the ivory statue he had made, prayed to Aphrodite, the goddess of love, and married her daughter, who had become a human being. The idea of the figure produced by Tagawa may not reach Aphrodite in heaven, but it has a certain power that shakes the hearts of many people on this earth and makes the people who see it sprout the "love."

田川 弘

Profile

1959 年、生於日本宮崎縣。畢業於大分縣立藝術短期大學附屬綠丘高等學校（美術班）後，進入名古屋藝術大學繪畫科（西洋繪畫）就讀。立志成為一名平面畫家。1988-93 年，獲獎中日展「大賞」、「佳作賞」（名古屋市博物館）。1995 年，邂逅林 浩己原型的擬真人物模型，1996 年第一次著手進行擬真女性人物模型的塗裝。之後於 2011 年開始以成為專業的人物模型塗裝師為目標進行各項活動。舉辦過多場個展。除了日本國內之外，田川 弘的粉絲支持者遍佈海外各國，是公認的擬真人物模型塗裝界的巨匠大師。

Born in Miyazaki Prefecture in 1959. He graduated from Midorigaoka High School attached to Oita Prefectural College of Arts and Culture and entered the Nagoya University of Arts, Department of Painting (Western painting). From 1988 to 93, Winning the Chunichi Exhibition "grand prize" and the "honorable mention award" (Nagoya City Museum). In 1995, encountered the figures works of Hiroki Hayashi and began painting the first "real woman" figure in 1996. Began his career as a figure painter in 2011. The master of the real figure painter holds many personal exhibitions and has many fans not only in Japan but also in many countries out side of Japan.

對於虛像的偏愛 Obsession with fictional image

外表看起來神聖不可侵犯，但又隱含了惡魔般的部分，呈現出一位少女成長為女人的過程，一個意志堅定的女性形像。這份要在畫布裡描繪出立體人物模型的堅持，至今仍歷久不變。

It is an image of a woman with a strong will in the transition from a girl to an adult. A demonic thing that is hidden with divinity. It is like drawing a figure on the canvas. His idea and attitude stays the same.

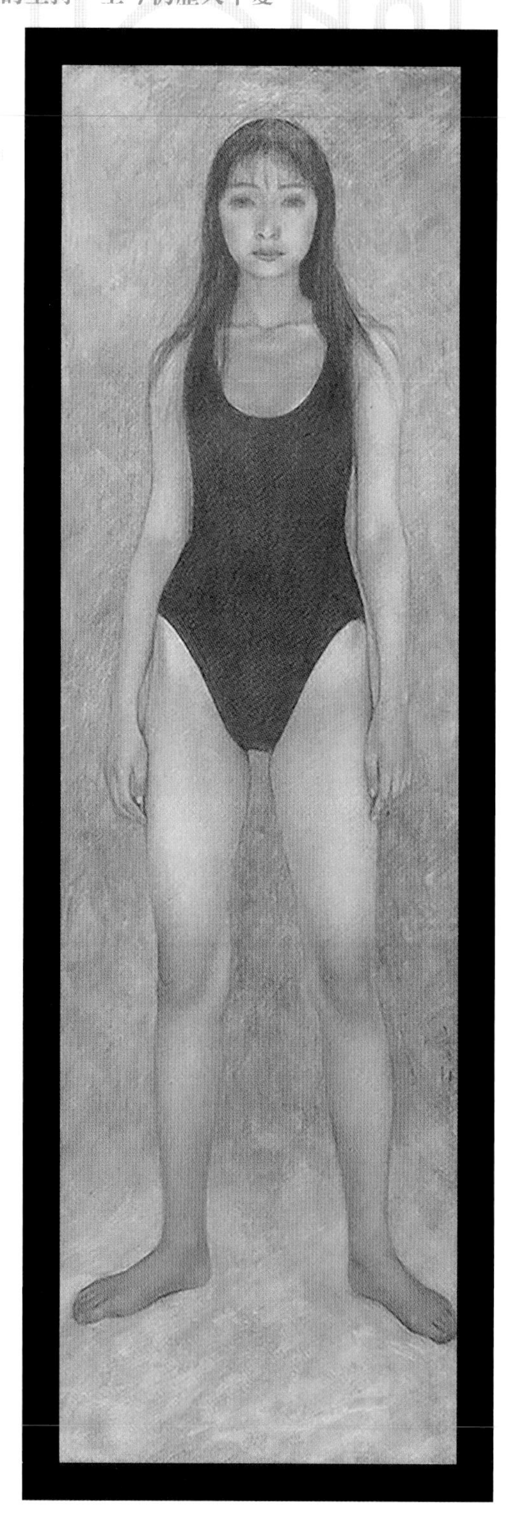

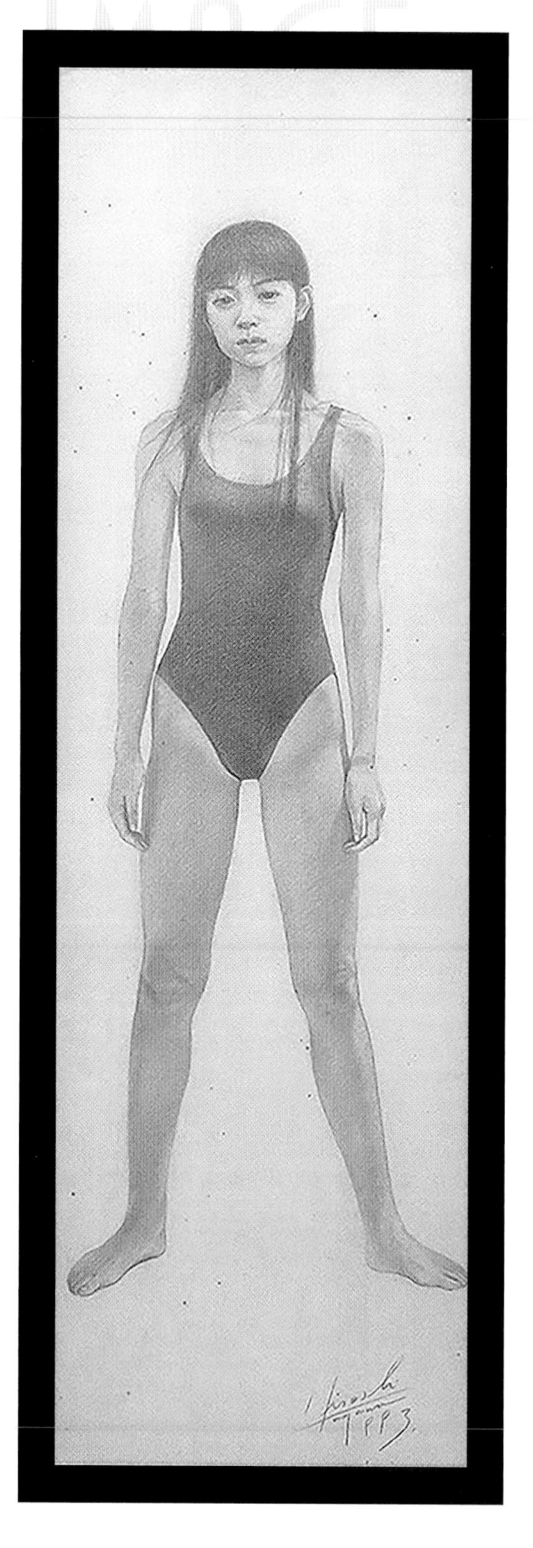

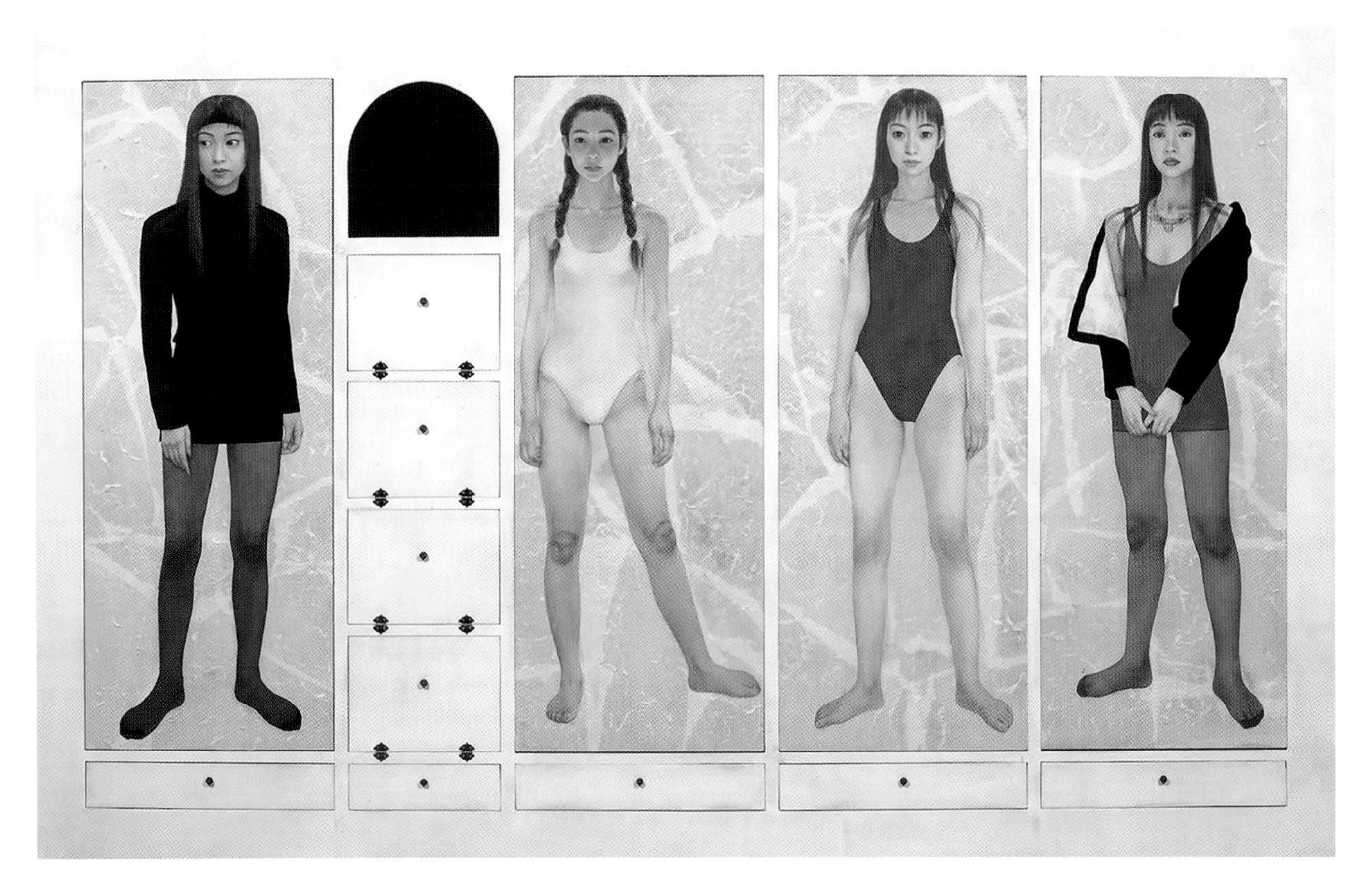

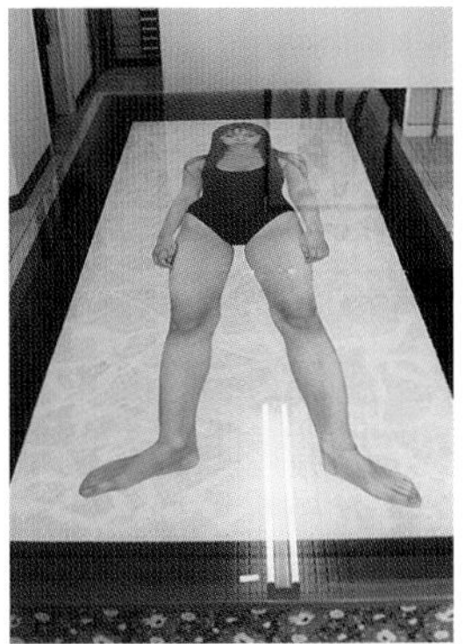

上面這張照片，是田川 弘過去舉辦個展時，用來迎接來會場參觀者的置物櫃；裝設在中央及下部的抽屜是可以拉出收起開關的構造。左側是以相同目的設置的桌子。像這樣對於不單純只是使用平面展示的藝廊空間形態的探索，直接連結到了田川 弘現在藉由立體與平面併用的作品展示方式。

In the image above, a drawer in the center and bottom opens and closes as a cabinet welcoming visitors to past exhibitions. On the left is a similar table. These approaches to gallery space, not just flat space, lead to the current exhibition of 3 dimensional and flat art.

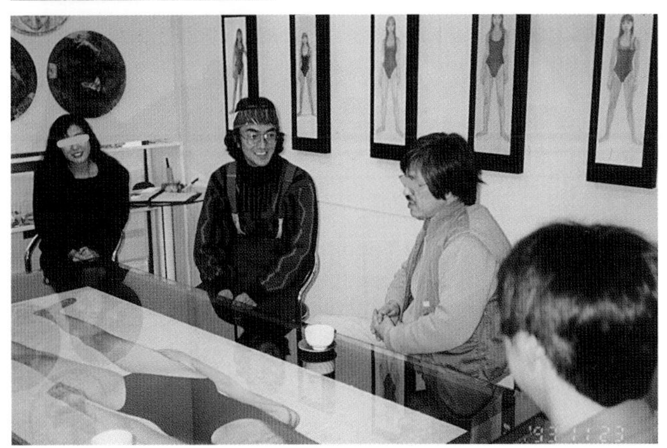

大多數的人，在從小到大的成長過程中，或多或少都會接觸到市松人形這類日本娃娃，或是法國娃娃、莉卡娃娃，以及布偶娃娃等各式各樣的人偶，並藉由對這些娃娃投入愛情的行為，學習到珍重寶貝某些事物的重要性。在其中還可以學習到雖然是虛擬的，但是與其他人相處的方式。伴隨著人的成長階段，有一天會從這種「代償行為」中畢業。但是對於人偶這種所謂「模仿真人形象」的物品，願意去格外注入愛情的人相信也不在少數。那只是單純對於美麗事物的喜愛？還是想要拿在手裡放在身邊的收藏欲望？又或者是當成託付愛戀之情的對象？每個人都各有屬於自己的故事。本篇介紹了田川 弘尚未開始著手製作立體作品前的平面作品，應該也可以解讀出田川 弘所「偏愛」的事物為何吧！

Since childhood, many people have come into contact with dolls, including French dolls, and stuffed animals. Through the act of loving them, they have learned the importance of treating things with care, as well as the pseudo way of interacting with others. Many people graduate from these "substitution act" as they grow up. There are quite a few people who show more than a little affection for dolls and other things that imitate the so-called "Human shape". It is simply a desire to love or keep something beautiful. There are as many feelings as there are people. The works introduced here are the two-dimensional works before Tagawa started to make solid figures. We can probably see Tagawa's "Obsession" from this work.

pygmalion©

WAKKER

2018 年 1/5 比例 樹脂鑄型套件
原型 klondike

2018 1/5 Scale Resin cast Kit Sculpted by klondike

這是作品風格的氣氛介於擬真與動漫角色人物兩者之間的女性造形作家 klondike 的原型作品。這次田川 弘在塗裝時思考設定的情節背景是一對每天晚上都在街頭遊蕩的美麗姐妹花 Tan 與 Kao。兩姐妹心中懷抱著遠大的夢想，有一天要成為這個街區知名的歌手，一對姐妹嘻哈歌手……。一邊在腦海中想像著這樣的情境，一邊動手製作。黑色的編織帽是將原始套件稍作改造而成。帽子上的標籤是用自己製作的水貼。閃閃發亮的項鍊是以飾品用的鏈條加上美甲裝飾用的小零件製作而成。

The original mold by Klondike, a female sculptor. Her mold has a characteristic of being situated between the real and fictional character. Tan and Kao are sisters who prowl the streets night after night. Their big dream is to making a name as a hip-hop musician。This story was created while building this figure. The black beanie is a modified kit parts with the self-made decal. Necklace is crafted with accessory chains and nail ornaments.

田川 弘塗裝不可或缺的工具與材料 Essential materials for Tagawa's paint work

由零件組裝到塗裝前的事前準備、正式塗裝，一直到最後修飾完成為止，身為塗裝師接連催生出令人驚異作品的田川 弘，掌中到底藏有什麼樣的工具與材料呢？

To produce unique works as a finisher, from the assembly of parts in preparation before painting and final painting, what kind of materials are in Tagawa's hands?

油畫顏料 Oil paint

塗裝的主要材料是 KUSAKABE 公司的油畫顏料。當年還是美術系學生的時候，使用的是 Holbein 好賓的顏料，不過因為有位尊敬的學長使用的是高價的 KUSAKABE 公司的油畫顏料，所以就從當年成了我夢想中的顏料。特徵是混色後也不容易變得混濁，主要使用的是發色良好的含鎘類顏料。此外，將鈦白色與群青色混色，經過長時間後，群青色會變得愈來愈淡。

The main material of the paint is oil paint by KUSAKABE. When he was an art student, he used Holbein's paint, but it was his dream to use the expensive oil paint of KUSAKABE that his senior whom he respected used. The main feature of this oil paint is that it is hard to get dull even when mixed with other colors. He mainly uses cadmium colors for its vividness. When titanium white and ultramarine are mixed, ultramarine will become thin due to aging.

打底用塗料 Primer (Undercoat paint)

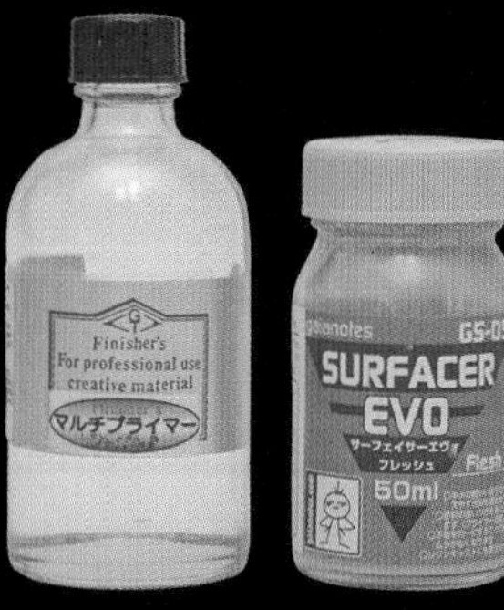

肌膚部分的打底塗料使用的是 Gaianotes Surface Evo Spray 的粉紅色，以及同公司的瓶裝 Surface Evo Fresh 塗料以空氣噴筆進行塗裝。其他部分則會視質感表現所需選擇塗料，基本上是使用 Soft 99 公司的灰色底漆補土噴罐來噴塗。

The skin area is airbrushed with Gaianote's Surface Evo Spray pink and Surface Evo fresh. The rest depends on the texture representation, but the soft 99 Corporation's gray surfacers pray.

紙調色盤 Paper Paint Pallet

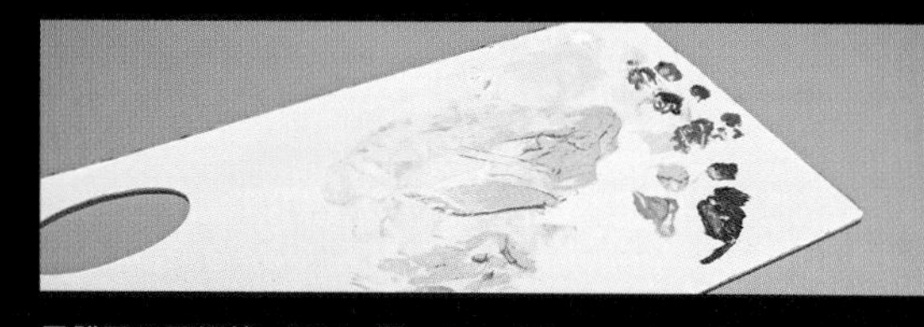

弄髒了只要撕掉，用下一張就能繼續使用，非常好用的工具。除了調合顏料之外，裁切假睫毛加工時也很容易辨視，相當方便。此外，如果油畫顏料使用後將整個調色盤用保鮮膜包起來放入冰箱的話，可以保存一個星期左右。

It's useful because they can turn it over when it gets dirty and use a new one. It is convenient as it is easy to see when cutting eyelashes other than paint. After using the oil paint, cover with plastic wrap and refrigerate for about a week. It is convenient as it is easy to see when cutting eyelashes. Cover the oil paint with plastic wrap and refrigerate after using to preserve it up to a week.

畫筆 Brush

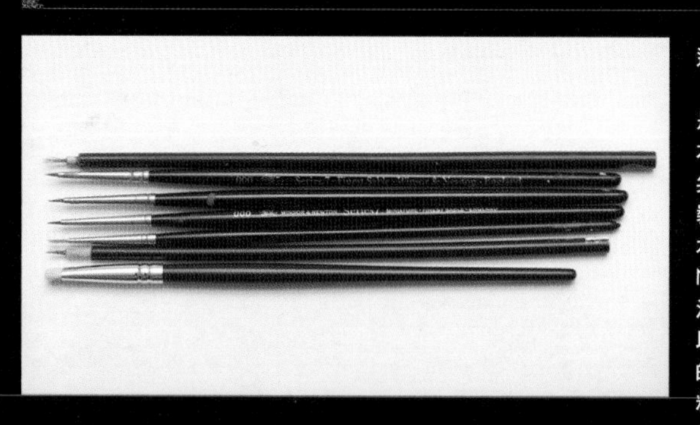

這裡使用的是 TOKYO.NAMURA 的黑貂毛畫筆（中）&（小），不管是臉部的大面積上色或是描繪血管等細節，都很容易調整筆壓，是日本繪畫也會使用的畫筆。特徵是韌性很好，筆尖的毛不容易散開。臉部及其他細節部分的塗裝，使用的是 WINSOR&NEWTON 溫莎牛頓水彩用畫筆系列 7 的 Miniature 微型畫筆 No.000。Miniature 微型畫筆的筆毛較短，一次沾附的顏料較少，也可以拿來使用在微縮模型以外的塗裝。特色是筆尖不易散開，以及柔韌的筆觸。這枝畫筆田川 弘未曾使用於硝基系顏料的塗裝，而是以油畫顏料為主，偶爾也會使用於琺瑯漆塗料。此外還有將畫筆前端剪短，專門用來戳染的畫筆；以及將筆尖拔到只剩一根毛，用來撈起假睫毛的面相筆等等，依照作業的需要而自行改造的各種不同畫筆。

Uses TOKYO NAMURA sable hair brush (medium) & (small), a brush for Japanese style painting, which makes it easy to adjust the pressure when drawing large areas of the face and blood vessels. It is characterized by good stiffness and keeping bristles in fine shape. For painting face and other small parts, Winsor & Newton Watercolor brushes Series 7 Miniature No. 000. Miniatures series have shorter tips and keeps less paint, so non-miniature versions are also used. The focal point of this brush is its good firmness and its shape.This brush has never been used for lacquer-based paint, and is used mainly for oil painting and rarely for enamel paint. There are also several types of brush that he made for the necessity steps, such as the brush that tip was cut for tapping, and fine brush that tip was cut into a single hair for painting eyelashes.

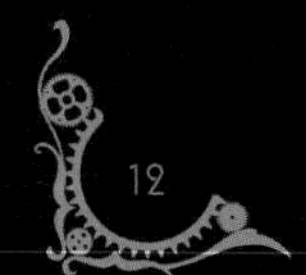

接著劑 Glue

這是 Horizing 公司的環氧樹脂類接著劑 Magic Smooth。接著強度足以因應 1/7 比例以下的小比例模型組裝，或是可以讓 1/6 比例以上的大尺寸模型不需要打上連接軸就能固定。45～60 分鐘即開始硬化，約 12 個小時後便可完全硬化。接下來是瞬間接著劑 CYANON 與 Gaianotes 的瞬間彩色補土 Fresh。CYANON 與彩色補土是用來填埋氣泡痕跡，或者是撫平表面凹凸不平。除此之外，若與粉狀的瞬間接著補土一併使用的話，還可以修補較大的高低落差，非常好用。如果再和硬化時間較長的 Magic Smooth 併用，也可有助於固定零件使用。

The manager of a famous model shop in Osaka recommended him the Magic Smooth, an epoxy adhesive made by Horizing. Use it for assembly of small scale models up to 1/7 scale and its bonding strength is strong enough to glue light weight parts of lager scale models without any reinforcement. Curing begins in 45 to 60 minutes and completes in about 12 hours. Instant glue Cyanon DW and Gaianotes Instant adhesive color putty fresh. Those are used for filling air bubbles and smoothing uneven surface. It is useful for correcting large gaps when used together with powder of instant adhesive putty. It is also useful for fixing parts by using it with Magic Smooth, which takes a long time to cure.

銼刀及筆刀 Files and Hobby Knives

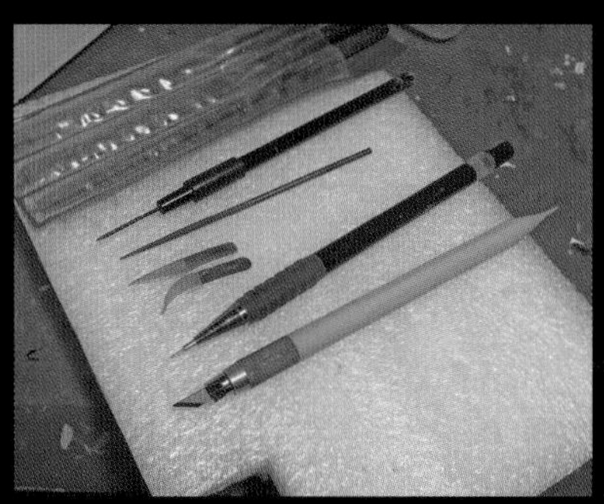

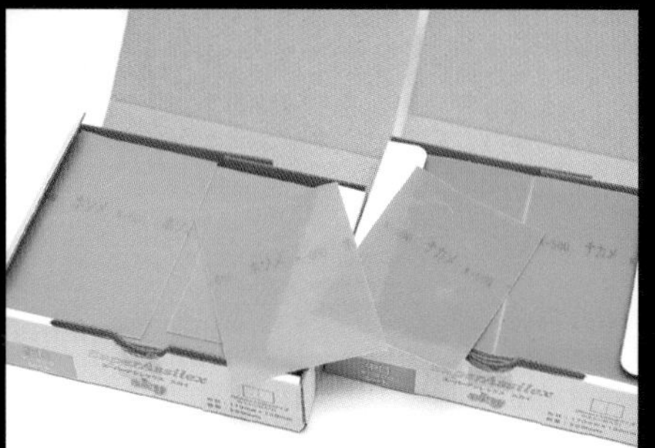

處理零件時，使用的是 OLFA 牌的美工刀，或者是 TAMIYA Craft Tool 田宮工具系列的基本銼刀（細目 Double Cut）。美工刀的刀刃只要稍有缺角就立即更換新品；若是銼刀棒的溝槽被碎屑阻塞了，就隨時以黃銅金屬刷來將碎屑刷除，徹底做好這樣的基本工具保養是很重要的重點。研磨砂布使用的是 Kovax Super Assilex 的 K-360/K-500/K-600/K-1000/K-1500 等各型號產品。相較於一般的耐水研磨砂紙，質地更柔軟，對於曲面的追隨性較高，也較耐用。墊在木塊上使用，或是用指腹按壓使用，以鑷子挾著使用都可以；或是裁切成小塊，黏貼在牙籤的前端，用來修飾細部細節。

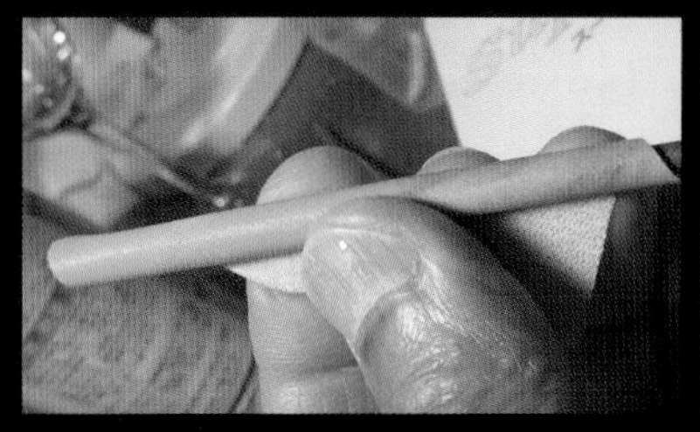

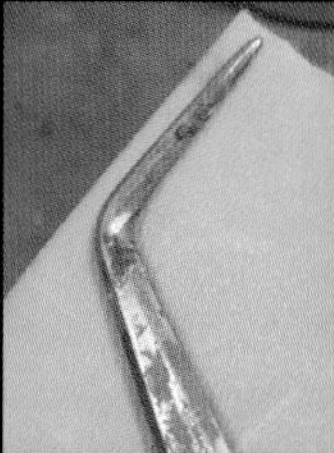

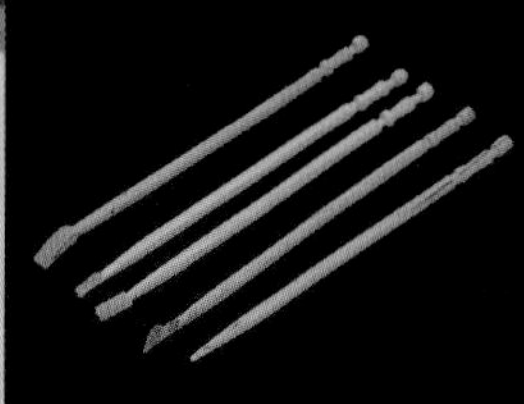

The parts are refined with OLFA Art Knife and Tamiya Craft Tool Series basic file set (Smooth Double-Cut). The point is to thoroughly carry out the basics such as replacing the blade of the art knife with a new one as soon as it chips, and clean the file with a brass brush frequently when the dregs are stuck in the groove.The fabric file uses Kovax' Super Assilex K -360/K -500/K -600/K -1000/K -1500. It is softer than general water-resistant paper file, has a better follow-up to curved surface, and last longer also. It can be attached to a wood block, pressed with the tip of a finger, pinched with tweezers, or cut into thin pieces and attached to the tip of a toothpick for finishing details.

保鮮膜 Plastic Wrap

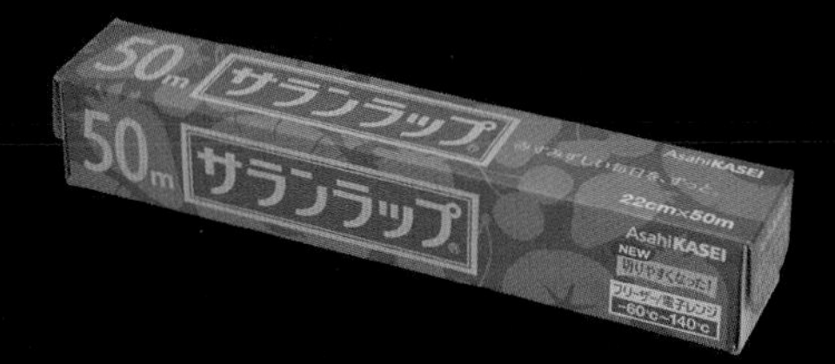

製作過程中，如果用徒手拿取套件的話，會讓手上的油脂沾在樹脂零件上。或者是不小心碰撞時，有可能造成塗料剝落。為了要避免這樣的狀況發生，所以要以保鮮膜來大範圍包覆保護零件。質地柔軟，自由度高，使用起來很方便，也可以用來保存紙調色盤上的顏料。

It is used to cover a wide surface for masking to prevent oil and grease sticking to parts during the production, and to prevent the paint from peeling off when it collide with other objects. It is flexible, easy to use and also useful for preserving the paper pallets.

其他塗料 Other Paints

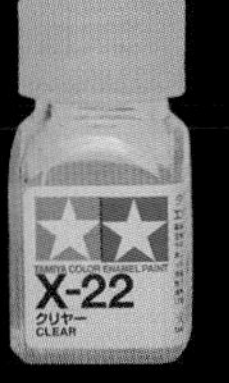

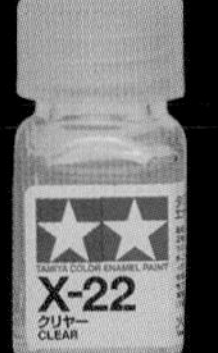

琺瑯漆塗料使用的是 Tamiya 田宮的產品。金屬、稜鏡塗裝是以 Gaianotes 的空氣噴筆進行噴塗。電鍍風格的塗裝則是將 Spazstix 的 Ultimate Mirror Chrome 塗料以空氣噴筆來噴塗。

Uses Tamiya enamel paint. He used a lot of clear paints with brush. Metal and Prism paints for airbrushing paint are from Gaianotes The metal plating tone paint is created by spraying Ultimate Mirror Chrome from SpazStix with airbrush.

顯微鏡 Microscope

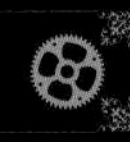

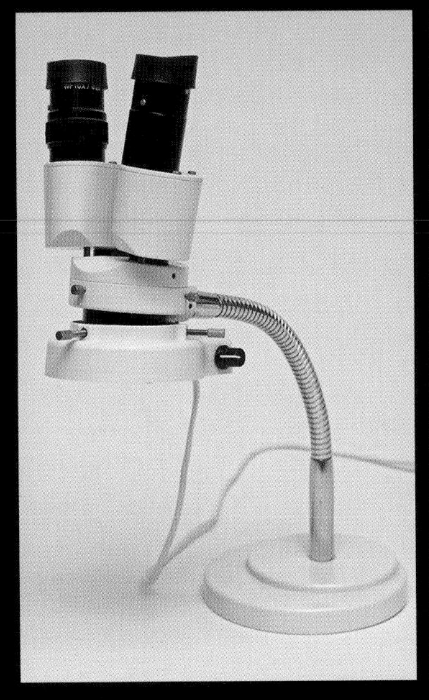

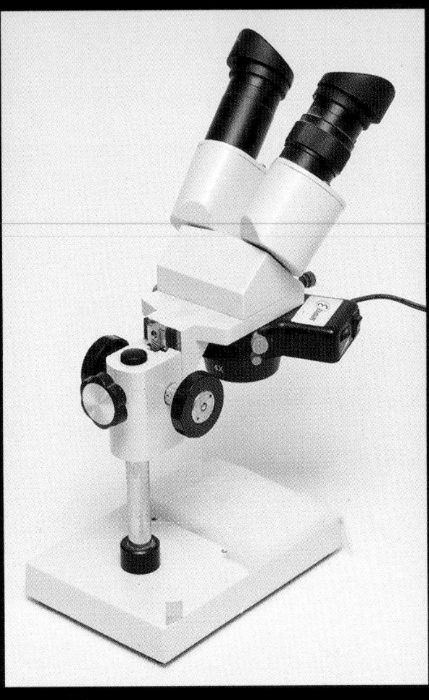

原本當初不認為有必要購買這個工具，但為了要將作品讀入 PC 電腦後製的事前確認作業，以及為了能夠更接近自己所憧憬 SNS 社群媒體上朋友發表的高解析度作品的品質，最後決定還是購買了顯微鏡。2011 年購買的是可以切換放大 20 倍與 40 倍的機種，目前也仍在第一線使用中。剛開始使用時，透過鏡頭觀看到的影像晃動得讓人頭暈，所以好一陣子都收在抽屜裡沒拿出來用。直到改採將對象物固定後再以鏡頭觀察的使用方法，才克服了頭暈的問題，開始了顯微鏡人生。作業時主要使用的是 20 倍的放大倍率，但因為視野會集中在一點上的關係，所以會和頭戴式放大鏡一起併用，不時還會以裸眼來觀察整體。後來再透過友人的介紹，2014 年購買了 8 倍的實體顯微鏡，算是用來填補上述 20 倍的實體顯微鏡與頭戴式放大鏡兩者之間的放大倍率。觀察對象的深度較深，可以讓我以手拿著對象物的方式進行作業。使用的頻率是 40 倍約 5%、20 倍約 50%、8 倍約 10%、頭戴式放大鏡約 35%。

At first, he didn't feel it was necessary, but decided to purchase it because he wanted to check the images on his PC and saw some high resolution works from SNS friends. In 2011, he bought a 20 and 40 switchable magnifications model, and he is still using it. When he first start using it, he was intoxicated with the image inside the lens and kept it in the closet for a while, but he got over it by fixing the object to stand and started working with the microscope. He mainly uses 20 magnifications, but since the viewpoint becomes too narrow, the work is done with combined use of a head magnifier glasses and checking the whole work with the naked eyes. A friend introduced him to a stereo microscope with 8 times magnifications which he purchased in 2014. The depth of field is so deep that you can work with the object in your hand. The frequency of use was 5% for 40 times, 50% for 20 times, 10% for 8 times, and 35% for the head magnifier glasses.

頭戴式放大鏡 Head Magnifier Glasses

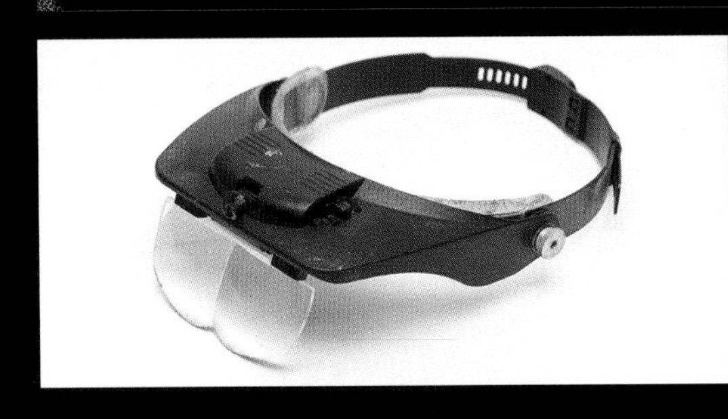

這是在 2009 年還沒有購買實體顯微鏡之前，作業時主要使用的 2.5 倍的頭戴式放大鏡。當時最小的模型製作尺寸是 1/20 比例的人物模型，描繪眼睛的瞳孔時，也只能在畫筆前端沾上顏料，透過這個頭戴式放大鏡，幾乎是依靠直覺進行描繪。即使現在已經使用實體顯微鏡作為主要工具，由於視野寬廣方便確認整體狀態的關係，仍然會有相當程度的頻率使用頭戴式放大鏡這個工具。

Before he started using stereo microscope, this 2.5 magnification head mount magnifier was mainly used. At the time, he was making 1/20 scale figures, but the pupils of the eyes were also colored with the tip of the brush, and he drew almost by instinct while using this head mount magnifier. Even now, with the introduction of stereo microscopes, it is easy to see the whole body from the wide field of view, and they are used quite frequently.

烘碗機 Dish Dryer

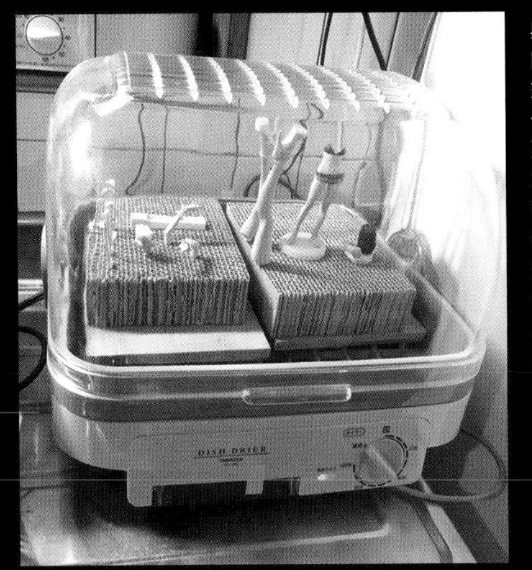

與一般模型使用的硝基漆塗料不同，對於慣用硬化時間較長的油畫顏料的田川 弘來說，烘碗機是一件不可或缺的工具。不論冬或夏，溫度都設定在 36～38 度，大部分的物件都是烘乾 1～2 天後即可進入下一道步驟。以使用頻率來說，基本上電源都沒有關掉的機會，但到目前為止使用都很正常沒有發生故障。

The dish dryer is indispensable for the Tagawa, who uses oil paints that take longer to harden than ordinary model lacquer paints. The temperature is set at 36 ~ 38°C from summer to winter and most of the time, he can move on to the next step in 1 to 2 days. He keeps the power on because of its frequent use.

※此非產品原本設計之用途，請依使用者個人責任判斷是否使用。

空氣噴槍 Airbrush

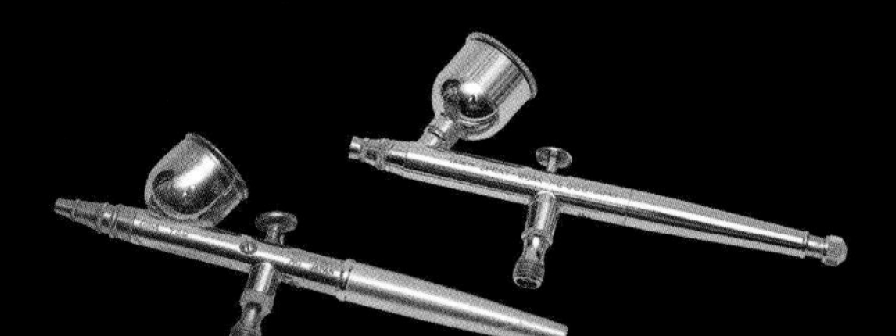

底漆補土或是珍珠光澤塗裝會使用 Creos 的 PS266 0.5mm 噴嘴。硝基系或鍍膜系塗料則使用 Tamiya 田宮的 Spray-work HG00G 0.3mm 噴嘴。壓克力塗料使用的是 Holbein 好賓的 Dash Y3D 0.3mm 與 Dash Y2D 0.2mm 噴嘴。2 液型 Urethane clear 聚氨酯透明高光漆塗裝使用的是中國製的噴嘴。本來已經做好便宜沒好貨的心理準備，沒想到卻意外的耐用。

Creos PS -266 0.5 ㎜ for surfacer and pearls. For the lacquer paint and metal plating paint, Tamiya SPRAY-WORK HG00G 0.3 ㎜ is used. Holbein Dash Y3D 0.3 ㎜ and Dash Y2D 0.2 ㎜ for acrylic paint. For the two-component urethane clear coating, he uses a no brand airbrush he bought it with the risk of low durability, but it is better than he expected.

虎鉗 Vise

這是在生活資材店買的一般桌上型虎鉗，我準備了兩座，放在我的副工作桌上。除了想要用兩隻手同時作業的時候，或是將正在接著中的零件彼此加壓固定等等，這些工具本來就設計好的用途之外，在塗裝的過程中，用來暫時夾住模型的支架，好讓雙手可以休息片刻，也是相當重要的功能。又或者是拍攝製作途中的模型照片時也可以使用。是非常便利好用的工具。

This is a common tabletop vise bought at a hardware store, and is installed in secondary work desks. In addition to using it for its intended purpose, such as when you want to work with both hands or use it to hold and crimp parts that are being bonded together. It is also useful to hold the model while taking a break during the painting process and take pictures of models in the process of production.

UV 紫外線燈 UV Light

以前使用的是筆型的紫外線燈，但因為照射範圍太狹窄，缺點很多。這是美甲用的紫外線燈，上面及兩側面都有 UV 紫外燈管，底部則裝設有鏡子，可以達到全方位的照射效果。附有 2 分鐘的定時功能。

Penlight type was used before, but its irradiation range was too narrow. This is for nails, and UV light is on the top and both sides. A mirror is on the bottom, so it can be irradiated evenly. Comes with a 2 minutes timer.

超音波洗淨機 Ultrasonic Cleaner

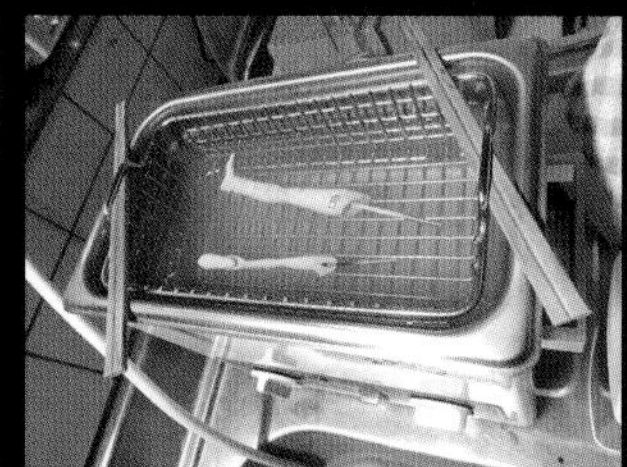

這是野澤製作所製作的超音波洗淨機 SD-B200H。因為需要長時間使用，而且一天當中會使用到好幾次，在同樣是塗裝師的友人介紹下，決定選擇工業用的機型，而非飾品用的廉價產品。洗淨液使用的是專用的洗淨液。貴金屬之類的物品只要放在中央清洗 10 秒左右已經能夠充分洗淨，但如果是清洗人物模型的話，則需要大約 5 分鐘左右。使用的訣竅是不要一次進行大量的洗淨。

Nozawa Seisakusho's SD-200H ultrasonic washer. Since he uses it for a long and many times a day, he chose an industrial model which is not cheap, but worth the price. Use a special detergent for the cleaning solution. It takes about 10 seconds to clean precious metals, but it takes about 5 minutes to clean figures. The key is not to wash too many parts at once.

濕毛巾 Wet Towel

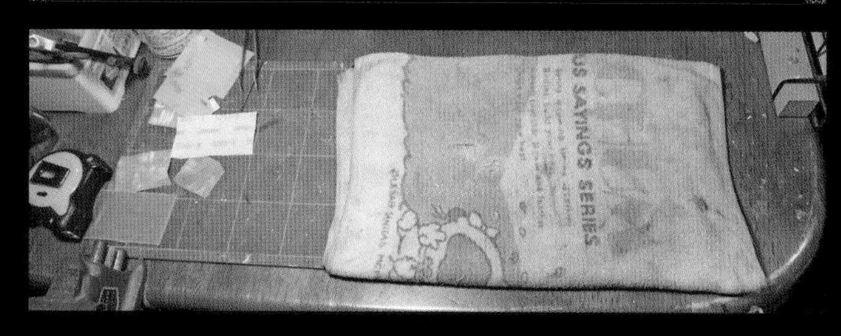

需要切削樹脂零件時，會在作業台上鋪設一條折疊起來的濕毛巾。

由於很少會需要大力的研磨處理，因此切削粉塵不會飄得到處都是，朝下掉落在桌上的粉塵則是會吸附在這條濕毛巾上。加上因為是折疊起來使用的關係，具備一定厚度，可以代替緩衝墊，發揮固定作業中零件的功能。作業後只要清洗過又能再次使用。照片中這條毛巾已經使用過 5 年了。

When the resin parts are shaved, a wet towel is folded and laid on the work table. He doesn't have to file too hard, so all the shavings will fall down and absorbed by the wet towel. It is thick because it is folded, and it is also useful as a cushion to hold parts during work. It can be washed and reused after work, and is already 5 years old.

假睫毛 Fake Eyelashes

製作 1/8 比例以上的尺寸，田川 弘一定會為人物植上睫毛。此時使用的是在價格均一商店即可買到的真人用假睫毛。將前端部分依喜好裁切成 0.4mm～1.2mm 來使用。市面上假睫毛的種類有很多，請選擇前端尖銳的產品。裁切時請使用美工筆刀等鋒利的刀刃。進行裁切加工時，如果放在紙調色盤這類方便確認的道具上，作業時切下來的假睫毛比較不容易不見。此外，要將假睫毛撈起來的時候，千萬不可以使用金屬鑷子這類堅硬的工具，會讓睫毛變得彎曲。正所謂「眼神能夠透露出千言萬語」，植毛加工這個步驟，可說是為人物模型注入靈魂的重要技法之一。此外也是能讓作品更加接近真人的「開光點睛」儀式。

Eyelash transplantation is always performed for sizes larger than 1/8 scale. These are fake eyelashes for people that are available at discount stores. Cut the tip to any length between 0.4 to 1.2mm. The key points in making false eyelashes are that although many products are available, a sharp tip product should be selected, a sharp blade knife must be used to cut the tips, and cutting on paper pallet will make the process much easier.. Also, when picking up the fake eyelashes, do not use something hard like a tweezers, it will be distorted. There is a saying that the eyes speak as much as the mouth, but this hair transplantation is exactly one of the techniques to put the soul into a figure and brings the work closer to people.

田川 弘的塗裝步驟 Tagawa's drawing procedure

傾注全心全靈在作品上的人物模型塗裝師田川 弘。從取得套件開始，透過了怎麼樣的步驟直到作品完成為止呢？讓我們來為各位解說概要吧!

The figure finisher Hiroshi Tagawa continues to produce art with his soul. Here is an overview of his process from obtaining the kit to completing the work.

STEP 1 購入 Purchase

GK 套件在市面上流通的件數有限。一旦看到自己喜歡的作品，請盡快確認庫存量，有貨就要馬上買下來。必須要有「喜歡的 GK 套件就算去借錢也要買下來」的心理覺悟。

Garage kits are limited in number. If there is anything you are interested in, check the stock immediately and purchase it.

STEP 2 開封 Unboxing

當套件送到後，馬上開封檢查是否有破損或是缺件。萬一有檢查到不良或缺件時，請立刻與廠商連絡，請對方因應處理。

Upon receiving the kit, I immediately opened the box and checked the parts (For damage or missing parts).

STEP 3 暫時組裝 Temporary Fitting

為了要掌握套件完成時的印象，先要將零件暫時組裝起來一次。除了預演塗裝的順序及組裝的方式這類實務作業步驟外，也要去「理解這個套件的背景」，思考這件作品所處的故事環境。

Temporarily assemble the parts in order to grasp the completed image of the kit. Considering painting order, method of construction and understand the kit.

STEP 4 洗淨 Cleaning

樹脂鑄型的 GK 套件大多數都含有二甲苯溶劑，也會有離型劑附著在表面。因此需要放入含有中性洗劑的熱水中煮過，使二甲苯揮發掉，也能夠去除離型劑。

Cleaning parts. Simmer in a pot to evaporate xylene, remove mold release agent and oil on the surface.

STEP 5 底層處理 Foundation Treatment

樹脂鑄型物在經過長時間後，會滲出微量的油脂。需要使用底漆或是補土底漆來先做好底層處理，隔開油脂。這裡使用的是米色的底漆補土。

The surface is coated with surfacer or primer to shut out the oil that leak out of the resin. Beige surfacer is used on this example.

STEP 6 肌膚的底層 Skin Foundation

使用油畫顏料，以面相筆在肌膚部分描繪血管。放入烘碗機烘乾 1～2 天，使油畫顏料硬化。接下來在身體各部位加上斑點。再放入烘碗機烘乾 1～2 天。

Using oil paint on the skin parts, draw blood vessels with a fine point brush. Put the parts in dryer for a day or two, then stain the body. Dry it for another a day or two.

STEP 7 開始正式塗裝 Start of Main Painting

進行第 1 層的肌膚塗裝。將身體塗上大致的色彩。從臉部開始作業，比較容易帶入情感。重點在於眉毛周邊以及髮際線的毛根表現。塗料的濃度要調整成能夠透出底層的血管及斑點的程度。

The First layer of skin paint. Put the color on the whole body. It's easier to empathize when you start with your face. The important thing is to express the roots of the hair, such as around the eyebrows and the hairline.

STEP 8 使油畫顏料硬化 Hardening the oil paint

使用烘碗機，促進油畫顏料硬化的速度。烘碗機的溫度設定為冬天 35 度～39 度，隨時保持電源開啟的狀態。透過使用烘碗機，不管是夏天或是冬天，都只要 1～2 日就可以硬化到能夠再上另一層色彩的程度。

Use a dish dryer to speed up the hardening of the oil. The dryer is always switched on in winter at 35 to 39 celcius. It will harden for overglaze in 1 to 2 days in both summer and winter.

STEP 9 第2層油畫顏料 The Second Layer of Oil

此時要將瞳孔和眼白、鼻孔、唇形、雙眼皮、下睫毛、眉毛等等的位置及大小決定下來。肌膚的部分要塗上稀薄如水的油畫顏料，以類似乾掃的技法來堆疊數層呈現出想要的色調。

The eye, nostrils, lips, eyes (Position and size of the white of the eyes), double eyelids, lower eyelashes, and eyebrows are started to be determined.

STEP 10 臉部的最後修飾 Finishing the Face

由於臉部的最後修飾需要反覆進行堆疊塗裝，所以要使用烘碗機來做油畫顏料的重點硬化處理。在等待硬化的過程中，開始進行衣服裝扮部分的第 1 層油畫顏料上色。

The oil paint laminated on the face is hardened.In the meantime, the first layer of oil painting on the costume begins.

STEP 11 睫毛的植毛 Eyelash Transplant

臉部的油畫顏料硬化後，接著就要進行田川 弘拿手絕技的睫毛植毛作業。先將市售的假睫毛前端部分裁短，種植在眼睛的邊緣。然後以 UV 硬化的透明樹脂接著固定。

Tagawa's specialty, eyelash transplantation. The fake eyelashes are cut short and planted at the edge of the eyes.

STEP 12 第3層油畫顏料 The Third Layer of Oil Paint

開始進行衣服裝扮部分第 2 層油畫顏料。在臉部的瞳孔等處塗上透明的琺瑯漆。頭髮的第 1 層油畫顏料也大約在這個時候開始著手進行。

The second layer of oil paint on clothes. Put enamel clear on the face. Start the oil painting of the hair.

※油畫顏料的顏色是藉由成分中的油分硬化後，達到乾燥、定著的效果。因此本書是將油畫顏料的乾燥狀態稱為「硬化」，但這與其他的硝基漆塗料或是琺瑯漆塗料的乾燥是相同的意思，並非有另外經過什麼特別的處理步驟。

STEP 13
修飾及第3層油畫顏料
Finish and oil color 3rd layer

放入烘碗機 1～2 天後，臉部和衣服裝扮的部分都會一併完成硬化。接著開始進行頭髮的第 2 層油畫顏料、同樣放入烘碗機烘乾 1～2 天。

In the dryer for 1 to 2 days to finish with your face and clothes. Start the second layer of Oil paint for the hair. After that, let it sit in the dryer for 1 to 2 days.

STEP 14
組裝 Assemble

放入烘碗機烘乾 2～4 天。輕輕碰觸表面，確認油畫顏料的硬化狀態。如果油畫顏料已經乾燥到可以作業的程度，就將各零件組裝起來。

Dry for 2 to 4 days in the dryer. When the oil paint has hardened enough to be touched lightly, the parts are assembled.

STEP 15
最後修飾 Final Finish

將人物模型安裝至事先準備好的底座。零件組裝時，即使小心翼翼地作業，仍有可能會有一部分塗裝出現剝落，此時要進行修補。並且觀察整體，如果發現到有不自然的地方，也一併進行修改。

Fixing to base and correcting color peeling due to the assembling. Also, look at the whole figure and correct anything strange.

STEP 16
完成立體作品
Completion of a 3d work

一邊飲用咖啡，一邊進行最終確認。其實這是田川弘最感到開心的時間。這個步驟不需要動手，是以類似「用眼神製作」的感覺進行。

Final check. it's the best time for Tagawa. This step doesn't require his hands. It is like "Building with eyes".

STEP 17
攝影 Photo Shoot

田川 弘的人物模型都是拍成平面作品後才算是真正的完成。以數位相機（單眼相機）進行攝影。每一件作品大約都會拍下 150 張到 400 張左右的照片。

Taking picture with digital camera. 150 to 400 pictures are taken for every work.

STEP 18
編輯 Editing

將照片讀入 PC 電腦，以 Photoshop 進行後製加工。裁剪刪除不需要的部分，將色調盡可能調整到實際的色調，並且一定要在照片上標記版權標示。

Captures images to a PC and develops them with Photoshop. Trimming and approximate the color to real thing and put credit the photo.

STEP 19
發表 Foundation processing

平面作品完成了。田川 弘為了要完成這幅平面作品，在創作主題的人物模型這個立體畫布上進行了「描繪」作業，而這就是田川 弘在創作上的行事作風。最後要將作品發表在 SNS 社群媒體，或是模型完成圖發表網站上。

The completion of the plane work. For this work, Tagawa "draws" the motif of the figure. It is posted on SNS on and a website for posting finished images of models.

田川老師的作品都是經過這些步驟完成的哦！由於受限篇幅無法發表全部過程，就讓我們介紹幾個主要步驟吧！
Tagawa's work is created with those steps! I can't cover everything, so I'm going to focus on the main points.
©Kow Yokoyama 2008

1/20 Ma.K. 女性整備士(B)
瑪蒂娜（マルティナ）技師

2013 年 1/20 比例 BRICK WORKS
樹脂鑄型套件 原型 林 浩己

2013 1/20 Scale BRICK WORKS Resin cast Kit
Sculpted by Hiroki Hayashi

田川 弘的臉部描繪解析

Analysis of Tagawa's Face Drawing.

田川 弘總是將「我不是在塗裝，我是在描繪一張臉」這句話掛在嘴上。作業的內容本身以模型製作的觀點來看雖然屬於塗裝的範疇，但田川 弘的作品就是能讓人感覺「不像塗裝」。

Tagawa says that it is not painting, but drawing a face. The process is painting from a model making point of view, but Tagawa's works do not make you feel "painted" by any means.

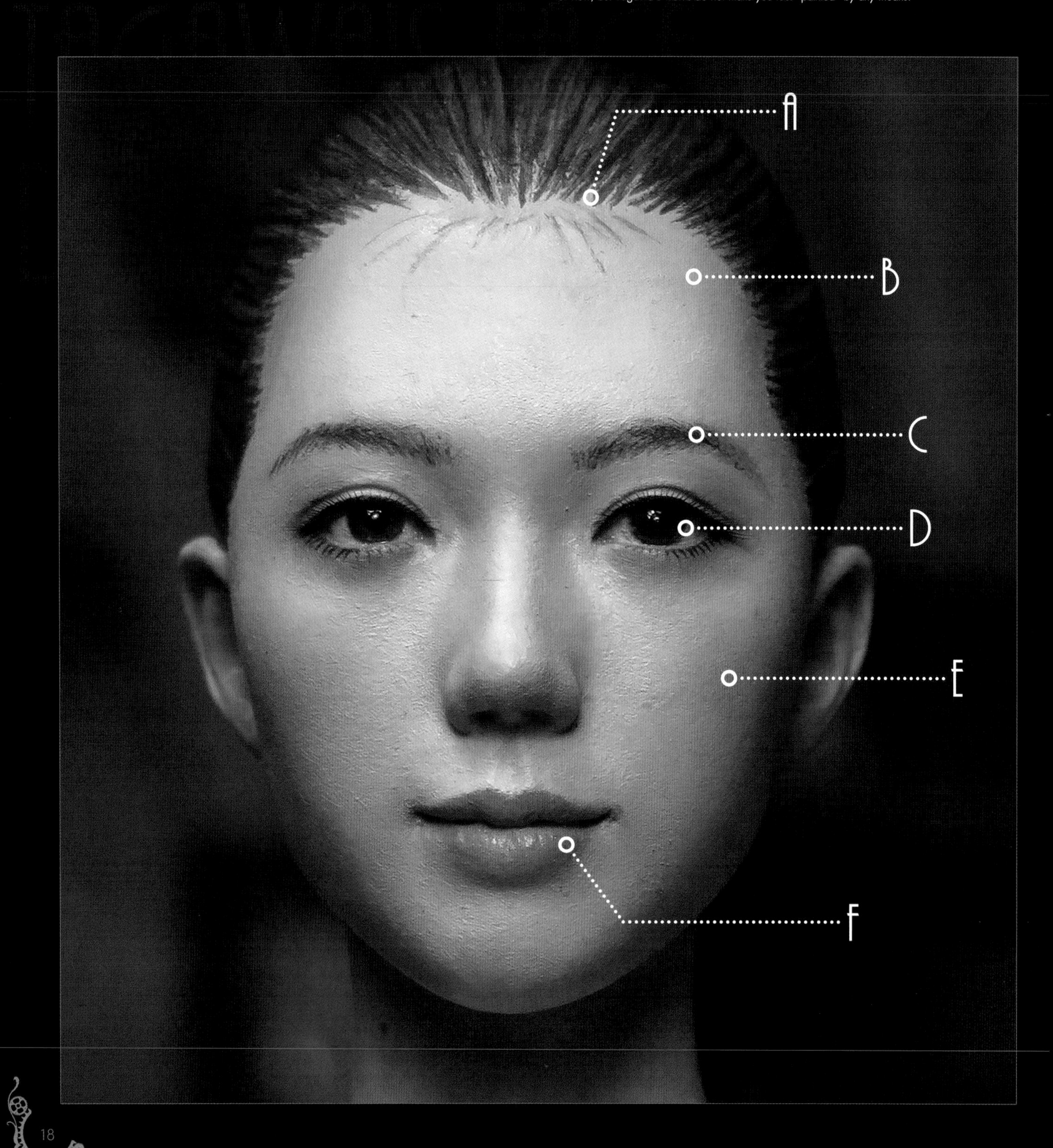

A 髮際線的毛髮

Hairline and Hair

範例作品是容易確認髮際線的馬尾造型。髮際線部分的製作重點在於皮膚上帶點藍色的色調表現，以及髮際線的頭髮長度所形成的額頭露出面積的微調整，再來就是鬢毛以及汗毛的細節呈現。另外別忘了髮際線部分的髮量較少，因此與其他頭髮層層相疊後的髮色相較之下，色調會顯得較更淡薄一些。絕對不是以單一顏色將整個頭髮平均塗布就完事了。此外鬢毛的表現會因為描繪的位置與大小不同，給人帶來不一樣的印象，會需要一些繪畫上的判斷力。

In this work, it is easy to check the hairline because of her pony tail. The key points here are the blueish tint of the skin, the fine adjustment of the volume of her forehead by the hair length at the hairline, and the drawing of loose and downy hair. In addition, the hair color and the hairline are less overlapping, so it should be noted that the color is lighter than other areas. It's not just a matter of painting in one color. Also,since the impression of loose hair changes depending on the position and size of the hair, an artistic touch is required.

B 肌膚的彈性

Firmness of Skin

如同嬰兒或少女般飽滿水嫩，而且充滿色澤、彈性的肌膚。田川 弘在描繪作品的肌膚時，基本上都是以這樣的狀態為印象進行作業。經過他的手中描繪出來的效果，即使描繪的是如病態般的白色肌膚，也完全不見絲毫的凝濁或暗沉，可以感受到由體內深處滿溢而出的生命力及躍動感，呈現出栩栩如生般的肌膚光澤。以肌膚的描繪來說，和色調同樣重要的就是光澤感了。也許很多人都會以為人物模型的塗裝都需要進行消光處理，但其實人的肌膚是比想像中的更具備光澤感。

When Tagawa paints the skin of his work, he generally has image of natural lustrous firm skin like a baby or girls.There is no cloudiness or dullness in even pathologic white skin, and the vitality and dynamism that overflow from the inside of the body,
Luster is just as important as color. Figure models tend to have a matte finish, but human skin is more shine than you think.

C 柔順的眉毛

Soft Eyebrows

眉毛的各項特徵，比方說形狀及長度、眉尾上揚或是下垂、濃密程度如何？色調又是如何？都能夠對人物的情感表現產生很大幅度的變化。基本上眉毛只是許多既細且柔軟的毛髮的集合體，但又要求在符合人物模型的角色個性與故事性的前提下，能夠藉由眉毛來呈現出表情的豐富變化。我們會需要使用較細的面相筆來將眉毛一根一根描繪出來，所以建議一開始先在紙上練習會比較好。眉毛的重要性與眼神相同，是能夠決定出最終表情的重要部位之一。

Eyebrows greatly change the expression of emotions depending on the shape, length, whether the eyebrows are raised or lowered, the thickness of the eyebrows, and the color. Basically, it is thought to be a collection of thin and soft hair, but it is necessary to change the expression of the figure in accordance with the character and story of the figure. You will have to draw one by one with a very fine point brush, so you should practice drawing on paper at first. Like their eyes, it's one of the key parts in determining the final expression.

D 會說話的眼睛

Volitional Eyes

在本件範例作品中，看起來是黑色的瞳孔，其實是由包含了綠色、藍色與橙色等不同顏色複雜交會而成。雖然這並非是以寫實擬真的技法來描繪呈現，但藉由透明塗料所形成具有透明感的重塗效果，賦予了瞳孔的立體深度。此外，覆蓋在整個瞳孔上的彩色透明漆所形成的光澤，讓人感受到了強烈的意志及生命力。眼睛的內眼角和眼尾的各處深邃陰影，以及自然反光映射在瞳孔上的高光，更進一步確認了眼神主人呈現出來的強烈意志。

Sable's eyes are expressed in this work, with green, blue and orange intertwined. It is not drawn with realistic paint techniques, but the transparent overcoat with clear paint gives depth to it, and the luster of the clear color covering the entire pupil gives it a strong will and vitality.

E 水嫩的臉頰

Moist Cheek

田川 弘想要表現的不是一般化妝在臉頰上呈現出來的腮紅效果。而是要體現出包含稚嫩、年輕、健康這些要素在內的紅潤肌膚，甚至還要有能夠讓人感受到溫暖體溫般的說服力。看過田川 弘的塗裝技法的人，大多會說出「好像女性的化妝」這樣的感想，但其實田川 弘本人並沒有自己在幫作品化妝的認知。倘若化妝這個行為的本來目的，就是要重現天然的肌膚原本所擁有的水嫩色調的話，以此思考便能夠理解田川 弘的理念了。

It is not just a cosmetic expression. The red color is persuasive enough to make you feel your body temperature slightly because it contains elements of youthfulness, and health. Those who knew the painting technique of Tagawa often said "the same as a woman's makeup" but he was not conscious of it. It makes sense to think that the makeup is originally an act to reproduce the color of fresh skin.

F 自然紅潤的嘴唇

Natural look of Lips

除非作品的設定是成年女性這類特定的情景之外，田川 弘不會使用畫上口紅的這種表現方式。嘴唇和臉頰、肌膚同樣是以自然的粉紅櫻花色與光澤感為追求的目標。不會有像口紅那樣界線明確的區隔塗色。帶著紅潤色彩的嘴唇，彷彿下一秒鐘就要張口對我們說話般栩栩如生。與稍微點綴在嘴角凹陷處的寂寥藍色形成對比，襯托出嘴唇的紅潤顏色。像這樣利用極端的對比色來襯托出美麗的手法，田川 弘將其稱為「美女與野獸」效果，除了應用在塗裝的領域之外，也在他的作品隨處可見同樣的手法。

He does not use lipstick expression except in certain situations such as adult women. There are the natural pink cherry blossom color and glossy feel on the lips as well as the cheeks and skin.There is no clear line like lipstick, but this lips makes you imagine that word are coming out from this reddish parts.A shade of blue slightly placed in the corner of the mouth, in contrast, accentuates the redness of the lips. Tagawa calls this technique of using such extreme contrast to enhance beauty the "Beauty and the Beast". This effect is not for just painting, but incorporated everywhere in his works.

跟隨田川 弘的作業步驟

Follow the steps of Tagawa's figure production.

從打開套件包裝、進行必要的底層處理、然後開始塗裝、一直到作品完成為止的一連串準備工作與作業步驟，就讓我們以照片解說的方式，一步一步為各位解開箇中訣竅。

After opening the box, we will explain the preparation and procedure from the necessary pretreatment to painting and finishing with photos.

底層處理 Foundation Treatment

市面上販售的組裝人物模型，大半是由 PU 聚氨酯樹脂製成的鑄型套件。在量產過程中，零件上很多會有為了從矽膠模具中脫模而塗上去的離型劑附著在表面。或者偶爾在脫模的時候，零件會出現因為受力不當而扭曲變形的情形。一般來說，會藉由將零件放入水中煮沸一定時間的方法，將表面的離型劑去除，或是修正零件的變形。不過離型劑只靠煮沸並沒有辦法有效去除，所以還需要經過另外的除去作業。

Most of the assembly figures on the market are resin kits made of urethane resin. In the manufacturing process, a release agent such as wax for taking out from the silicon mold is often adhered to the surface of the part. The mold release agent can also distort or deform the part during mold release, so it is common practice to boil the part for a period of time to remove the mold release agent and correct the deformation of the part. However, since the release agent is difficult to remove by boiling alone, removal work in a separate process is also required.

※作業中請充分進行換氣

1 煮沸這個步驟，除了修正樹脂鑄型套件特有的變形狀態之外，還有讓樹脂材料含有的二甲苯溶劑成分揮發掉的目的在內。這部分的處理方式，田川 弘也是經過不斷的收集資訊，進行試錯之後，才總算找到正確的方法。處理的訣竅在於不要將零件直接放入已經煮滾了的熱水，而是要由常溫水的狀態開始加熱。當然，我們需要準備一個模型作業專用的鍋子。

In addition to the purpose of correcting the deformation peculiar to the resin kit, boiling is carried out in order to remove xylene which is a solvent component contained in the resin. This is part of a trial and error process that involves constantly searching for the right answer.Instead of putting the parts directly into the boiling water, start with the water at room temperature. Naturally, a special pot for model work is prepared.

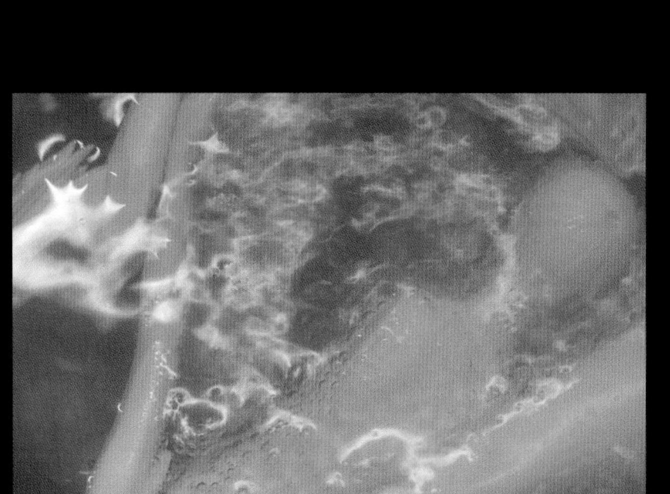

2 先正常加熱，直到水開始沸騰的溫度。依照零件的尺寸不同，再考慮到零件如果有變形需要修正的情形，最好是保持沸騰的狀態一段時間，好讓熱量能夠傳達到零件的內部。田川 弘建議的煮沸時間大約是 20 分鐘。鍋中的零件需要不停地攪拌。比較細小的零件受熱後會容易變形，所以攪拌時要充分的小心注意。

Bring the water to a boil. It depends on the size, but it is desirable that the heat is transferred to the inside of the parts to correct the deformation of the parts, so boil it for a while. Tagawa takes about 20 minutes for this process. The parts in the pot are constantly stirred. Thin parts are easily deformed by heat, so be careful when stirring.

3 再準備一個更大的鍋子，裝水加熱至沸騰後，把火關掉。將先前煮沸的零件倒進照片右方的大鍋中。需要注意的事項是因為零件在熱水中會變得柔軟，小心不要被其他零件壓迫到而產生變形。當所有的零件都移至大鍋之後，靜置等待鍋中的熱水慢慢冷卻到常溫為止。

Bring the water to a boil in a larger pot, turn off the heat, and transfer the all the heated parts to the larger pot. Keep in mind that the parts are softened by boiling water, so don't let them deform under pressure from other parts. Once all the parts have been transferred, wait until the water has cooled to room temperature.

4 仔細確認套件的澆口和分模線位置，進行打磨。依成型方法的不同，有時候會有微小氣泡混入樹脂當中，要將氣泡的缺口小心填埋補平。使用的工具是美工刀、筆刀、細目的金屬銼刀或研磨砂布。抹刀是用來填埋氣泡，以及修正零件之間的高低落差時使用。此外，零件要預先裝上金屬線，方便塗裝時拿取零件。

Carefully check the sprue and parting line of the kit and file those down. In some cases, small bubbles may have been mixed in with the mold, in such cases, make sure to fill up carefully. Use cutters, hobby knives, fine metal and cloth files. Spatula is used to fill air bubbles and correct uneven parts. Metal wires are attached to the parts to be used as handles for painting.

5 將所有零件暫時組裝起來，確認零件之間的吻合程度。此時也要確認塗裝時彼此是否會出現干擾，並將塗裝的順序決定下來。而且此時也是田川 弘遠遠地眺望著這件人物模型，在腦海中想像著屬於她的故事的一段「愉悅的時間」。然而不管人物模型的造形實力有多麼完美，有時候也會發生因為故事背景一直無法成形，導致遲遲無法開始進行塗裝作業的事態。

Temporarily assemble the parts and check the fit of the each parts. It's also a good time to check the painting process. It's a moment of fun when you look at the figures and wonder about her story. However, Tagawa said even for stunning shaped figures, if the story is not settled at this point, painting may not start.

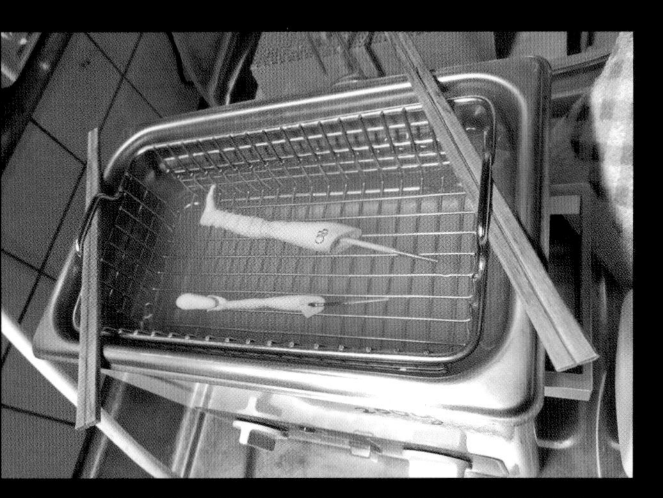

6 將暫時組裝好的零件再次拆開，重新插上方便拿取零件的金屬線，放入超音波洗淨機。請注意不要一口氣放入過多的零件。因為洗淨液使用的是工業製品專用的洗劑，所以在清洗完成後要用牙刷來仔細將零件表面的殘液刷洗乾淨。照片中的免洗筷是為了保持裝有零件的鐵籠（網子）不會沉入清洗槽底部而另外架設的。

The temporarily assembled parts are separated once, and the handle is attached again, and goes in to the ultrasonic cleaner. Be careful not to put too many parts at once. Because this cleaner is an industrial specification product, a special cleaning solution is used. After cleaning, the surface of the parts is carefully washed with a toothbrush. The chopsticks in the picture are for holding the net with the parts up in the air.

7 放置 1 天，等水分完全乾燥後，使用零件清潔噴劑或是硝基漆稀釋液來進行零件的脫脂處理。接著要馬上以底漆補土將整個零件噴上一層保護層，隔絕樹脂零件後續慢慢滲出外表的油脂成分。左側照片的脫蠟劑是為了要在噴塗保護層後，還需要進行打磨作業時，如果有樹脂材料裸露出來的話，用來去除油脂使用。和稀釋液或是零件清潔噴劑相較之下，脫蠟劑對於底漆補土的傷害較小，很適合使用於微修正的作業。

Leave for 1 day, and after the water has completely evaporated, degrease the parts with part cleaner or lacquer thinner.Immediately coat the entire surface with a surfacer to contain the oil that ooze out after degreasing. The parts cleaner spray is easy to use, but it has high volatility, so be careful about ventilation. The image on the left shows wax remover. It is used when the base of the resin comes out by filing after surfacer coat. It is more gentle to the surface than the thinner, so it is suitable for fine correction.

臉部的塗裝 Face Painting

臉部的塗裝相較於身體其他部位來說，步驟稍微有些複雜。此外，人物模型的塗裝一般都說可以參考女性的化妝方式，但田川 弘並非以成年女性的化妝為意象，不如說是以重現幼童或少女「不需要依靠化妝的肌膚彈性及光澤」為目標，將源自體內的生命力具現化呈現出來。不過並不只是單純流露出年輕的活力，而是將少女忐忑不安的心緒透過下眼瞼的黑眼圈以及病態般的蒼白膚色表現出來。

The painting process of the face is a little more complicated than that of other body parts. It is generally said that the painting of figures should be based on the makeup of women, but Tagawa's image is not the makeup of adult women, but rather the reproduction of infants or girls' skin, which does not require makeup to embody the vitality of them. However, it is not simply an expression of youth, but also expresses the dark circles on the lower eyelids and the pathologic whiteness of the skin as if it evokes the wavering feelings of a girl.

1 零件的底層處理完成後，進入塗裝作業的第一個步驟是將整個零件噴上底漆補土。所使用的塗料是 Gaianotes 的粉紅色噴罐式 Surface Evo Spray。噴塗時要注意盡量保持均勻，但又要避免塗層過厚。這裡的粉紅色會形成後續肌膚塗裝的底層陰影部分的色調。

After finishing the base treatment of the resin parts, spray the entire surface at the beginning of the painting process. Gaianotes' can spray, Surfacer evo pink is used. Spray evenly, but be careful not to spray it too thick. This pink color constitutes the shadow color of the base of the skin coating.

2 粉紅色的底漆補土充分乾燥後，再噴塗一層米色的底漆補土所使用的塗料是 Gaianotes 的 Surface Evo Fresh。目前因為噴罐式產品已經停止生產的關係，所以使用的是瓶裝的塗料。將空氣噴筆傾斜 45 度角，由上方進行底漆補土的噴塗作業。作業時要去意識到零件的凹凸形狀。

When the pink surfacer is dry enough, spray the beige surfacer. This is the Gaianotes' surfacer evo fresh. Bottled paint is used because can spray is no longer in production. Use an airbrush to blow the surfacer from above at an angle of 45 degrees to accentuate the unevenness of the surface

3 噴塗至底層的粉紅色像是由內側向外浮現出來的感覺即告完成。待其乾燥後，與身體部分的肌膚塗裝一樣，加上斑點的表現。使用的塗料是將淺鎘紅色與淺洋紅色兩色混合後的油畫顏料。將顏料以溶劑油稀釋成濃度接近清水，但帶有一些黏稠感的程度，再使用筆毛較硬的畫筆沾附顏料後，用手指的指腹將顏料噴刷彈色在零件表面。

Complete the spray as if pink is coming out from the lower layer. When it gets dry, it adds a stain just like the skin paint on the body. It uses a mixture of cadmium red light and light magenta. Do not dilute the paint with petrol too thin. The paint is placed on the tip of a hard brush and is sprinkled on with the tips of your fingers.

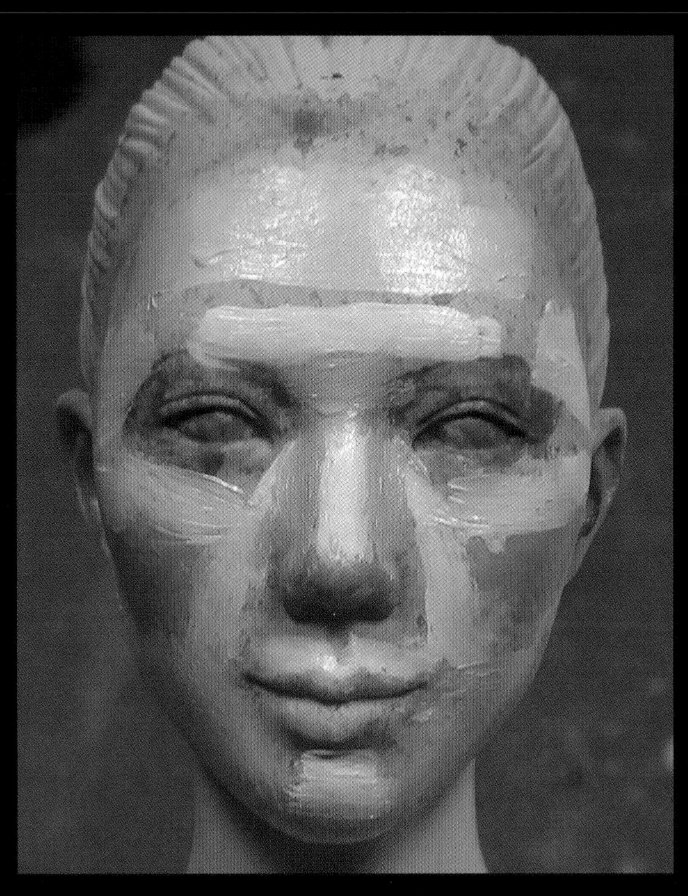

4 當斑點的塗裝充分乾燥後，接著在臉部的各個部位以基本的油畫顏料進行塗色。由於肌膚的底色已經塗裝完成的關係，接下來就要正式進入描繪臉部的步驟了。構成肌膚顏色的油畫顏料基本色，使用的是鈦白色、淺鎘紅色、淺鎘黃色、檸檬鎘黃色、鎘綠色以及群青色這些顏料的混色。另外，關於動漫角色的人物模型塗裝的基本概念，因為原本的平面插畫作品的陰影部分，是以較深的顏色來描繪，所以在凹陷形狀的部分一般會使用陰影色來上色。不過田川 弘並沒有採用這種在陰影部分使用比基本色更深的顏色來上色的手法。如果曾經使用過油畫顏料這個媒材的話，可能就比較能夠理解，田川 弘因為過去描繪的平面作品都是具象畫的關係，不只是在陰影的部分，而是在所有的面塊中，都可以觀察得出更多樣的色彩，所以將這些色彩呈現在作品中是再自然不過的事情。關節的彎曲部分，因為皮膚被拉長的關係，皮膚厚度會變得較薄，使得皮膚底下顏色較深的肌肉組織會透出皮膚表面，呈現出紅色調。同時，被拉伸的皮膚色調也會變成比原本的皮膚色更淡的顏色。一邊要像這樣去理解肉體的構造，同時要將觀察的對象從實際的肌膚之外，擴大到照片、寫實插畫、具象畫等等，進而培養出到底有哪些顏色混合成哪種色調的觀察眼力，並能夠將其落實應用在作品上的表現技法。田川作品的本質，無法僅用「擬真」一句話解釋，更可以說是將「繪畫天份」具現化之後的結果。

After the stain is sufficiently dried, a basic oil color is applied to each part of the face. The base of the skin is already painted, so it's time to start drawing her face.The oil used to create the basic skin color is a mixture of titanium white, cadmium red light, cadmium yellow light, cadmium yellow lemon, cadmium green and ultramarine.As a basic concept, shadow parts in the original illustration for character figure are drawn in a darker color, so it is common to place shadow colors on the concave parts of the figure. However, Tagawa did not adopt such a method for expressing shadow part of the figure. It is easy to understand for those who have experience in oil painting, but since Tagawa has drawn figurative painting, he naturally feels a lot of colors not only in the shadow areas, but in every aspect, and it is natural to incorporate them into the expression of his work.At the joints, the skin becomes thinner because the skin is pulled and reddish because the dark color of the underlying muscle tissue can be seen through the skin. At the same time, the skin is pulled and becomes lighter in color than it should be. While understanding such physical structure, it is a technique to express what kind of color can be seen not only in the actual skin, but also in photographs, realistic illustrations, and figurative painting. The essence of Tagawa's works, which cannot be expressed simply as realistic, but it is the realization of his artistic taste.

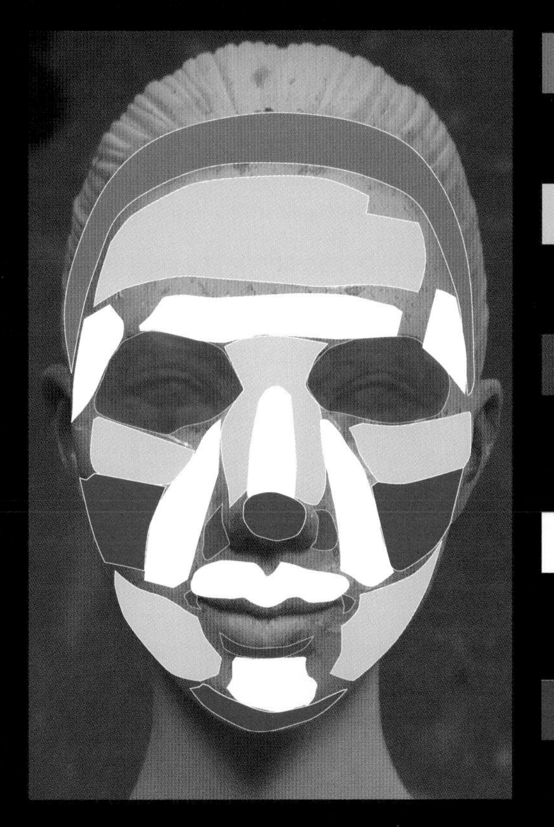

髮際線的部分，要意識到皮膚底下頭髮的毛根，在肌膚的基本色上混入少許鎘綠色油畫顏料後進行上色。這就像是刮完鬍子後留下的青色痕跡，這麼形容可能比較方便大家理解。不只是額頭部分，太陽穴、後頸部以及耳朵後面這些頭髮生長的邊際都不要漏掉以這樣的手法呈現出來。

For the hairline, a mixture of the basic color of the skin with a little bit of cadmium green is painted, being aware that the root of the hair is under the skin. It is easy to understand if you imagine the blueish skin after shaving. Not only the forehead, but also the temples, the nape of the neck, the back of the ears and other hairline areas should be covered.

臉部肌膚的明亮部分的表現方式，是使用與身體肌膚相同的肌膚基本色，實際所使用的油畫顏料種類請參考上述顏料。當然，依作品不同，所使用的肌膚基本色混合比例也都各有不同。此外，還會以作品使用的基本色為基底色，在需要呈現出紅色調的部分另外調色使用。

The basic color of the skin, which is also used on the skin of the body, is used to express the bright part of the skin. It is made of a mixture of titanium white and cadmium red light. Needless to say, all the basic skin colors have different mixing ratios depending on the work. Based on this basic color, color for reddish parts are also made.

眼睛的周圍、眼頭、鼻翼的兩側、嘴角等凹陷得較深的部分，以油畫顏料的群青色上色。雖然乍看之下好像用色非常不自然，但在油畫、具象畫的表現手法中，在陰影的部分加入各種顏色反而是再一般也不過的技法。對於田川 弘來說，陰影裡面看起來是帶著藍色調。這種技法所呈現出來的效果，只要看看其他具象畫或是寫實插畫的作品就能夠一目了然。

The Ultramarine blue is used for around the eyes, inner corners of the eyes, sides of the nostrils and corners of the mouth. . At first glance, it may seem like the use of heavy colors, but in oil painting and figurative painting, it is common to put many colors in the shadow areas, and for Tagawa, blue is visible in the shadow areas. The effect is obvious in concrete and real illustration works.

臉部受到自然光線照射到的地方，也就是所謂肌膚色的高光部分，使用油畫顏料的鈦白色來描繪呈現。眉毛所在位置的眼窩上緣，以及頭部兩邊外側的側面、鼻梁、臉頰靠近鼻子的面塊，人中的兩側、下顎的正面等等都屬於高光的面塊。但是請注意這裡與關節彎曲部分等皮膚被拉伸部分的表現方式並不相同。

A titanium white oil solution is placed on a part of the face which is naturally exposed to light, or a so-called skin-colored highlight. Other highlight places are the upper part of the eye socket with eyebrows and its outer side, the bridge of the nose, the face near the nose of the cheeks, both sides of the human body, and the front of the jaw. Note that this is not a representation of the joints or other areas where the skins are pulled.

臉頰是以少許的淺鎘紅色與淺洋紅色油畫顏料，加上微量肌膚基本色混色後調製而成顏色來上色。鼻頭、額頭、下顎前端則是以前述的顏色，再稍微多添加一些肌膚的基本色進行混色後增加變化。雖然只是細微的變化，但可以看得出臉頰部分會稍微呈現出紅色調較多的狀態。

The cheeks are a mixture of cadmium red light,little bit of light magenta and a tiny amount of basic skin color. The forehead and top of the nose are changed by mixing the above colors with more basic skin colors. It can be seen that there is a slight reddish tinge on the cheeks.

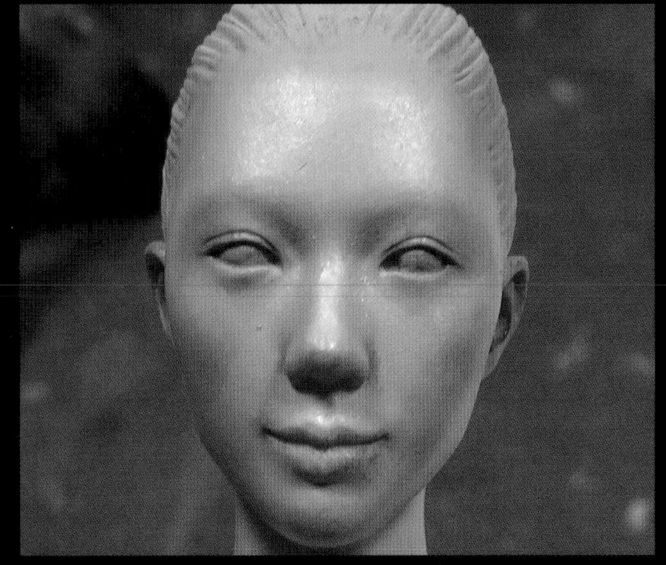

5 整個上色過一次後，使用將筆尖前端裁切整齊，增加接觸面積的自製畫筆，以戳染的方式來使整體色彩的邊界彼此暈染融合。首先要從額頭這類面積較寬廣的位置開始進行。

Once the colors have been applied, use a hand-made brush that has been cut flat at the tip to increase the contact area. First, start with a relatively flat area such as the forehead.

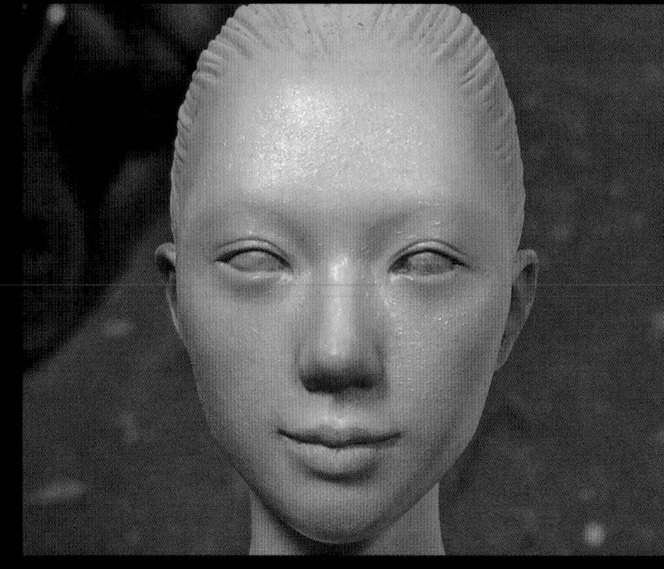

6 臉頰及鼻頭、下巴等處同樣以自製的戳染筆來讓油畫顏料彼此暈染融合。可以看到整個臉部的塊面顏色逐漸暈染融合到自然的狀態。剛開始覺得很突兀的藍色也漸漸融入周圍。作業時，不要所有的位置都進行相同的暈染融合處理，以免過於呆板。一邊作業時要一邊思考哪些部分要保留下來。

In the same way, tap the cheeks, nose, and chin to blend the oil color. You can see that the face is gradually becoming more natural. The blue color, which was originally distinct, gradually blends in. Instead of just blending to the flat surface, work on it while considering which parts to leave the color.

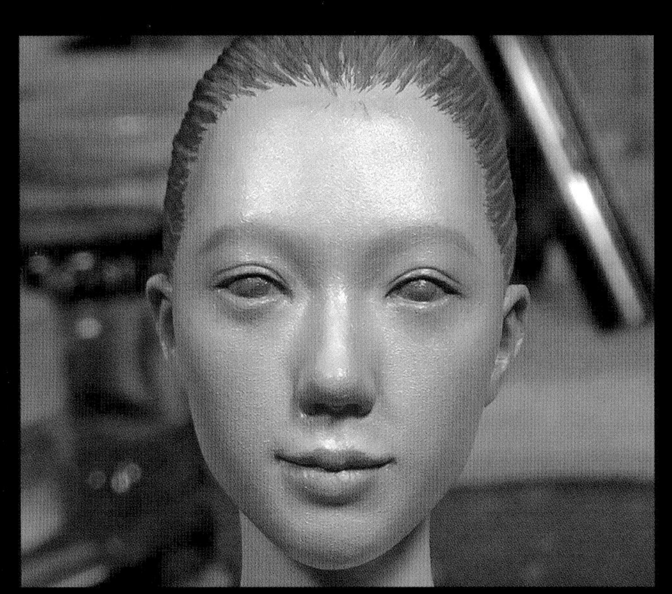

7 基本的油畫顏料都暈染融合處理完成後，接著要開始描繪頭髮和眉毛。頭髮的描繪作業請參考另外的步驟內容說明。這裡只要先確認頭髮的髮尖要描繪到哪個位置為止，同時確認整體的比例平衡即可。決定好髮際線後，接著再開始描繪鬢毛和汗毛。漸漸地，人物的表情就開始浮現出來。然後再把眉毛的「位置參考線」標示出來後進行描繪。

After blending the basic oil paint, draw hair and eyebrows. As for the hair paint process, please refer to another section, and we will proceed while checking the balance such as the drawing length of the hair ends. When the hairline is decided, start to draw loose hair and downy hair. Gradually the expression is appearing. Next, draw around the eyebrows.

8 頭髮的塗裝確定之後，經過充分的乾燥時間，再一次回到臉部的描繪作業。這個步驟常常會需要拿在手上作業，所以使用保護膠帶先將頭髮的部分遮蓋起來。此時要將眼白的部分以及瞳孔的大小確定下來。嘴唇上雖然也會塗色，但重點在於並非是塗上口紅的概念。需要找出最自然的色調，並將其呈現出來。

Once the hair has been painted, allow enough time for drying before returning to the drawing process. In this process, it is necessary to cover the hair with masking tape. The white of the eyes and the size of the eyes are determined here. Also start to put some color on lips but it is important not to paint them like applying lipstick. You must find a natural color.

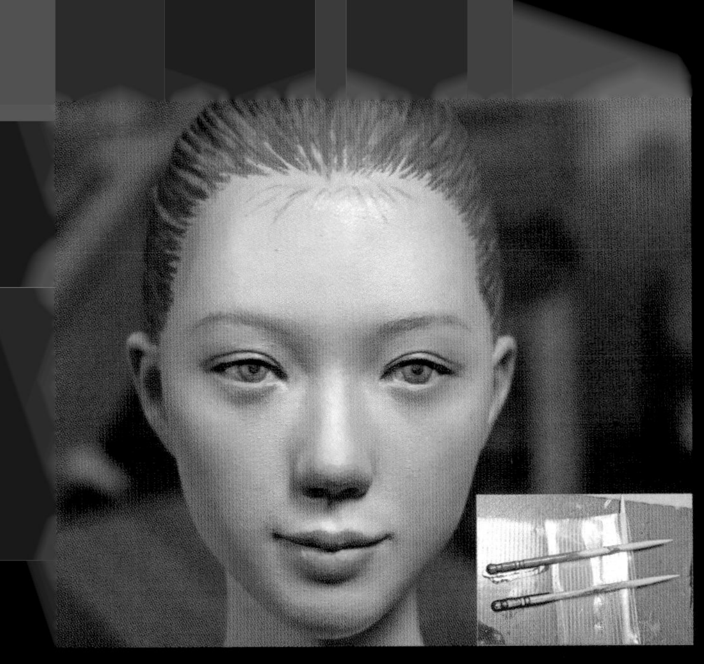

9 將雙眼皮呈現出來，並為整個眼睛的輪廓和下眼瞼塗上顏色，調整眼睛形狀，並畫出下睫毛。下睫毛並非植毛。將眼睛的輪廓和虹膜也描繪出來。在此步驟中，要將視線確定下來，所以需要格外慎重的細心作業。田川 弘的臉部塗裝作業看起來與成年女性的化妝過程相似，但本質不同。化妝這個過程的目的，實際上就是要重現如同嬰兒或少女般沒有暗沉、充滿彈性的水嫩肌膚。而田川 弘所描繪的臉部塗裝，正是要呈現出這樣的感覺。

The upper eyelashes are transplanted at this time, but please refer to the other page section for details. Tagawa's face painting process looks similar to women's makeup, but it is different in nature. The process of making up is, in fact, to reproduce the healthy skin of a baby or a girl, and the face painted by Tagawa is nothing but this expression.

10 我們很難在已經完全硬化的油畫顏料上再畫出極細的線條。如果不添加溶劑混合，畫筆將無法順暢運筆；但如果添加了太多溶劑，塗料又會在塗裝表面一下子擴散開來，無法描繪出極細的線條。但有一個簡單的方法可以達到這個目的。只要先塗上一層透明的琺瑯漆，再用面相筆沾取添加適量溶劑的油畫顏料進行描繪即可。這樣就能增加與塗裝表面的暈染融合程度，達到可以描繪出漂亮的極細線條的效果。這張照片是虹膜描繪完成後的狀態。

The pupils are further coated with UV curing transparent resin when the coated enamel clear hardens. The transparent material creates a deep finish. The highlights of the eyes in the image are not from paint, but from natural reflections of the resin. the resin is one of the useful material to express the shine of eyes.

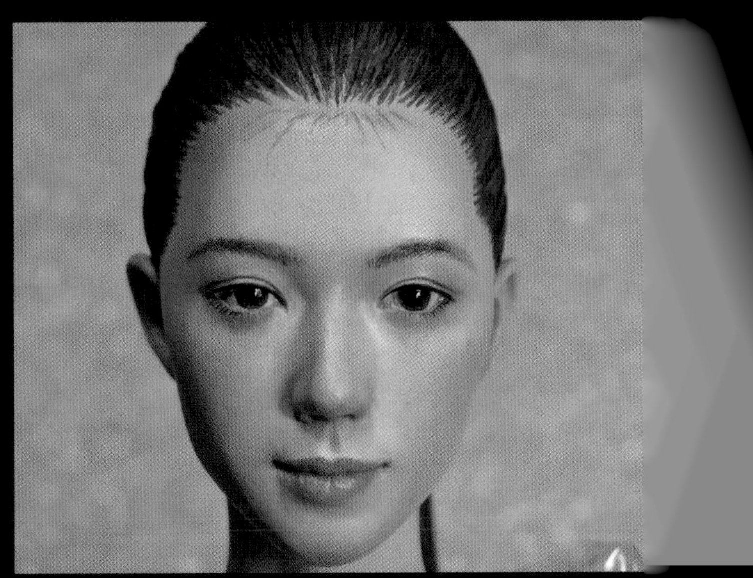

11 當細節描繪完成的虹膜硬化之後，將透明橙色漆與透明綠漆以 6:4 的比例混色，再與透明保護漆以 1:1 混合後，塗在眼睛上作為保護層。底層的藍色會透出表層，呈現出淡褐色的色調。當眼睛塗層的透明琺瑯漆硬化之後，再以 UV 硬化的透明樹脂加上一層保護層。因為素材是透明的關係，完成之後會使眼睛呈現出立體深度。此時接著就要開始睫毛的植毛作業

12 部塗裝完成了。臉頰的紅潤感、嘴唇的自然水潤感以及肌膚的光澤都展現出了人物的生命力。在意識到這其實是 1/8 比例，尺寸僅有數公分的臉部模型，便再一次讓人不禁感到驚訝。動漫角色類的人物模型往往會在眼睛以描繪的方式加上高光，但像本作這樣整個眼睛本身就能反射出光澤感，就不難理解為何田川 弘的作品總是能夠呈現出自然的視線了。

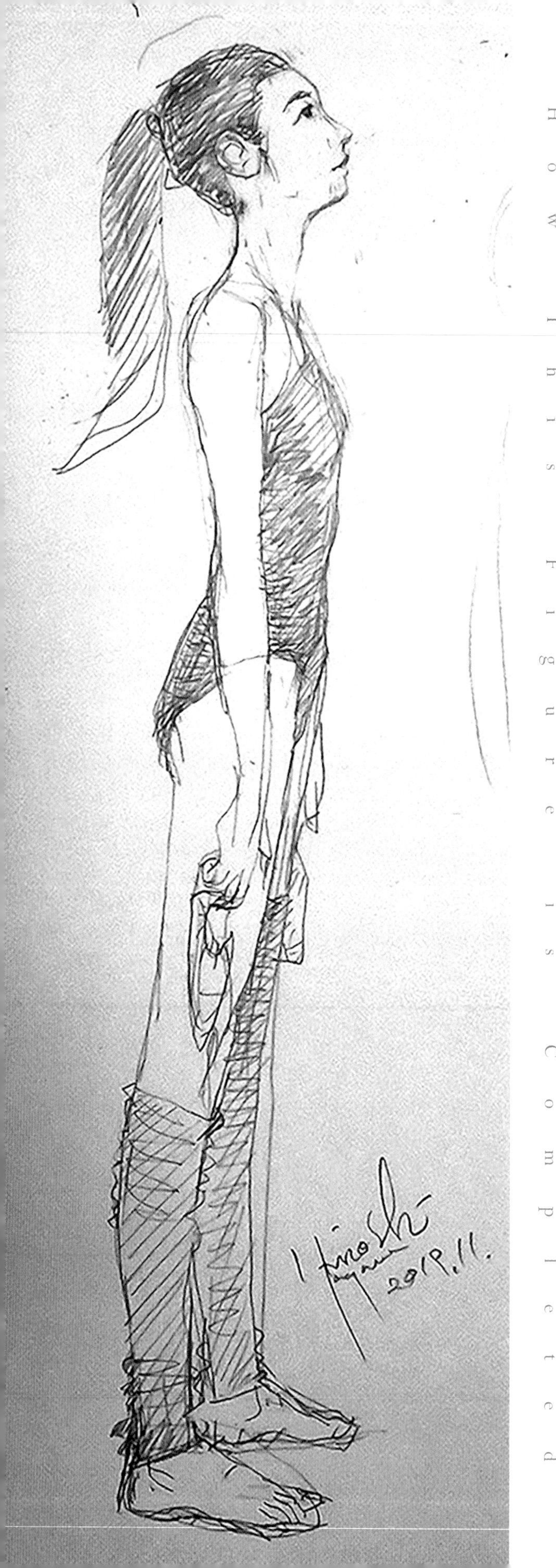

這件人物模型是如何設計出來的？

How This Figure is Completed

林 浩己如今已經開始使用數位工具製作原型。首先，依照田川 弘的要求，將雛型的檔案製作出來，然後再以此為基礎逐漸加上細節。

Mr. Hayashi is currently making original mold in a digital environment. First of all, we create the data as a model according to the order from Tagawa, and we work on the details based on this.

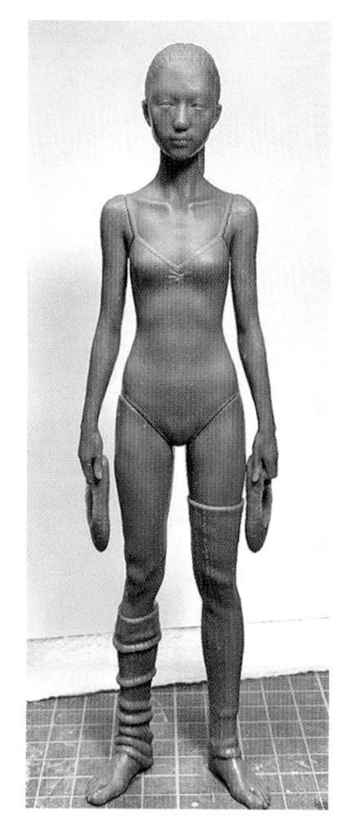

調整人物的頸部角度、雙腳張開的程度，並反覆調整因姿勢變化而帶來的重心位置變化。當姿勢確定下來後，先列印輸出一次確認實體。

Until the pose is determined, the center of gravity, the angle of neck and legs are repeatedly changed. Output the data once the pose is set.

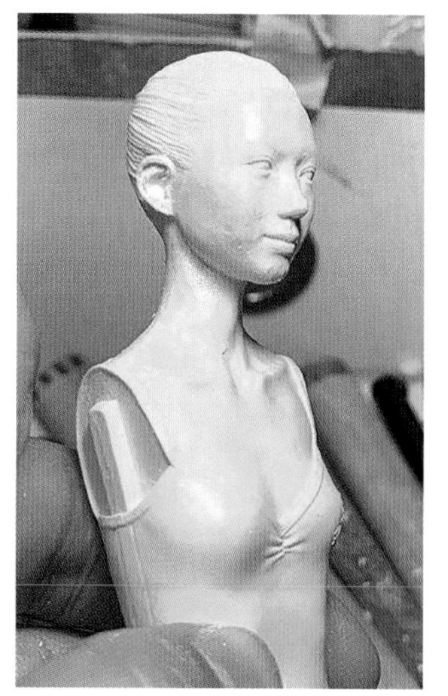

當所有的設計都確定下來後，林 浩己的作風是會將素材更換為環氧樹脂補土，好為細部細節做微妙的最後修飾。

After all the designs have been finalized, Hayashi replaces the material with polyester putty to create subtle details.

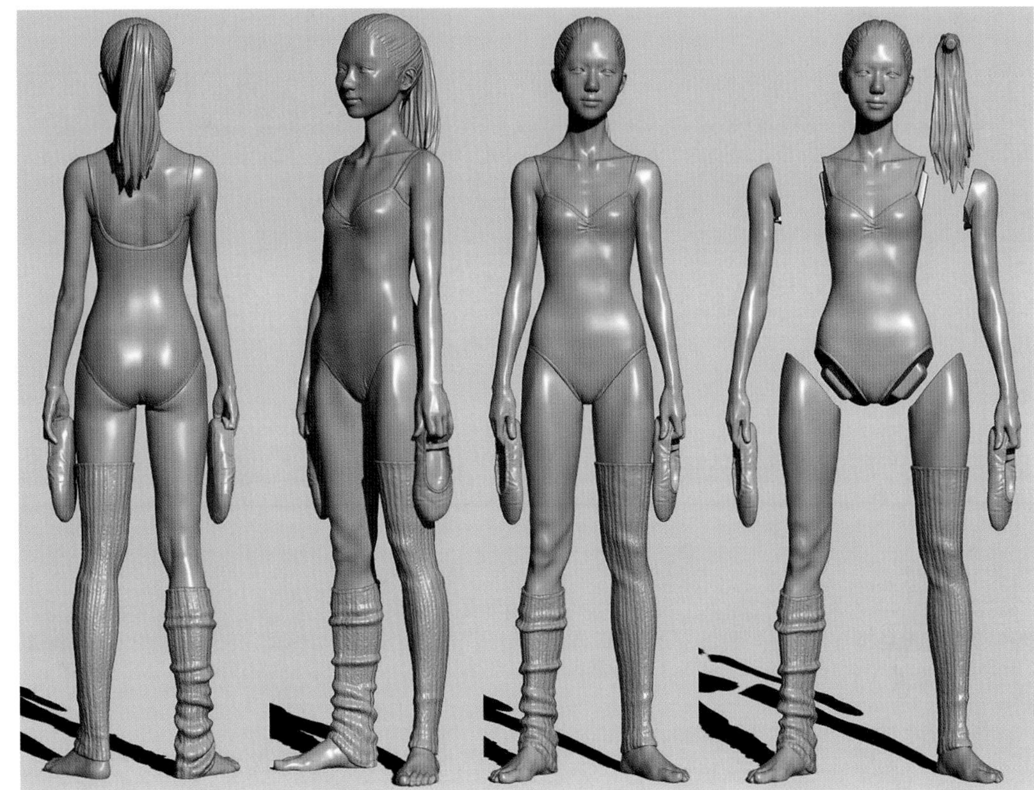

腳尖向外張開的程度，以及額頭的角度等姿勢調整；舞衣的款式設計以及馬尾髮型的氣氛；兩隻手裡拿著的練習用舞鞋、襪套等細部細節，所有環節都設計完成後的狀態。考量到適合後續商品化量產時的拆件位置也要在此時定案。將臉部原本略帶俯視的角度抬高到朝向正面，並讓表情浮現一絲微笑的感覺。塗裝師田川 弘全心全意提出的要求；原型師林 浩己全心全意地將其完成了。

All the necessary details, the opening of the toes, the angle of the face, leotard design, ponytail, training shoes held in both hands, and socks were fixed Suitable parts design for mass production are also performed at this point. Head angle was changed to look up to the front and smiling expression was added. Mr.Hayashi's response to the Finisher Tagawa's order was completed.

2020 年 1/8 比例 Atelier iT 樹脂鑄型套件 原型 林 浩己

2020 1/8 Scale Atelierit Resin cast Kit Sculpted by Hiroki Hayashi

Youth girl Extra edition

田川 弘為了完成這本著作，特地向人物模型原型師林 浩己委託製作了這件全新設計的人物模型。設計的主題是芭蕾舞者。但角色的身份並非當家的首席舞者，而是一位心中懷抱著各種煩惱，拚了命不斷練習的少女。設計的靈感來自於林 浩己原創的人物模型品牌 Atelier iT 的青春少女系列。希望能夠符合這個系列作品風格，以這個前提開啟了整個製作計畫。體型的設計是身材纖細、身高 170cm 左右、臉小、手腳細長。體型參考了 13～17 歲的俄羅斯女孩，以日本女孩來說大約相當於 16～20 歲左右。人物模型的姿勢原案是由田川 弘所設計。對於原型製作的作業，除了參考既有的音樂影片中登場的少女舞者身處的場景環境之外，也請教了芭蕾舞者提供實際經驗的意見。兩人都分享了所有的影像、圖像資料給對方，得以在不產生磨擦的情形下建立起設計方向的共識。近年來，林 浩己在製作人物模型時，也開始與田川 弘同樣會先去思考人物的故事背景。這次的關於人物的具體故事背景是由田川 弘先提出，再由兩人互相討論補強後，經過彼此都確認之後，才決定下來。雖然人物看起來只是單純的站立姿勢，但每位觀看者卻都能感受到其中的故事性……兩人合力的造形作品就是擁有這麼神奇的力量。

For this book's publication, Tagawa asked figure sculptor Hiromi Hayashi to make a brand new figure.The motif is a ballerina. However, she is not the so-called prima donna, but a girl with many worries and training hard everyday. The order was to make the same type of figure with energy Tagawa felt from Atelier it, an original figure brand developed by Hayashi, "Seishun Girls series" figure.

為作品植上假睫毛，幾乎可說是田川 弘作品的代名詞了。假睫毛裝飾在描繪精緻的眼睛上，呈現出只依靠塗裝絕對無法達到的真實感。那麼接下來就讓我們為各位徹底解說這個能夠讓人物模型更接近真人的手法吧。

Tagawa's specialty, eyelash transplantation. It dresses carefully drawn eyes and cann reproduced only by painting. Let us show you how he make the figure closer to the h with a thorough explanation.

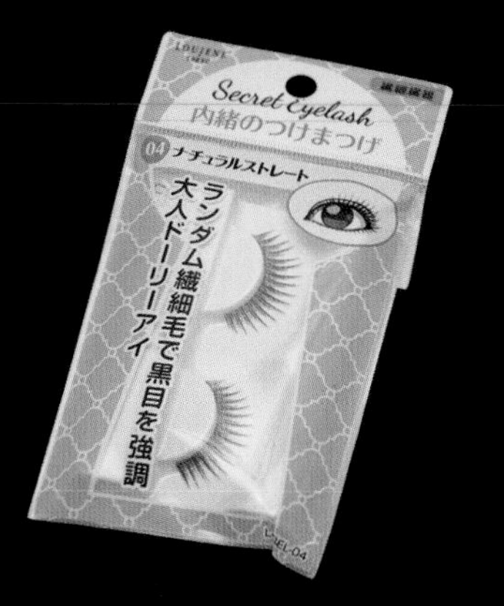

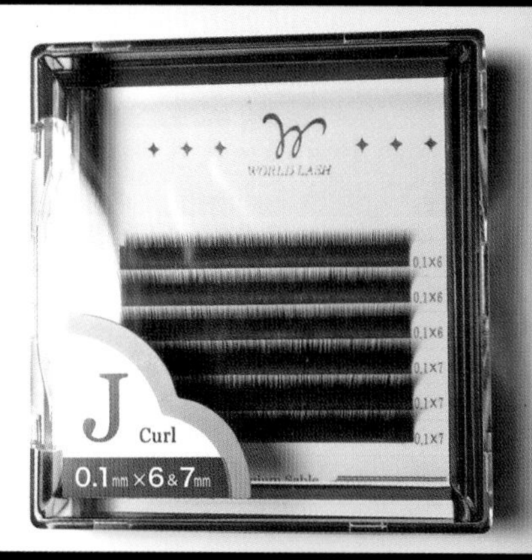

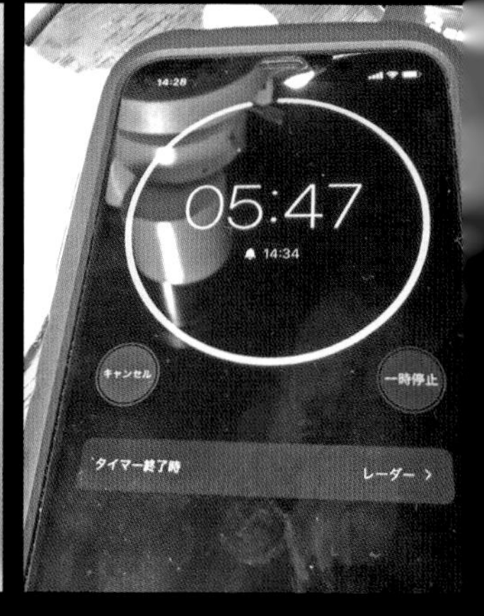

1 為人物模型植上假睫毛……相信很多人都曾經想過要這麼做。但是實際去付諸行將其視為一門技術學習，並且能夠讓作品在修飾完成後全無違和感的人，恐怕就不多了。必需要事先告訴各位，這是一件非常需要耐心的作業，但絕對不是不可能達到的領所使用的材料是在價格均一商店就能夠買到的真人用假睫毛。配合人物模型的比例尺寸，將毛的前端裁切成 0.4～1.2mm 的長度。植毛作業所需要的道具，首先是假睫毛本身。請選用前端尖銳的產品。實體顯微鏡要使用倍率 20 倍的型號。畢竟光用肉眼作業實在太過於困難UV 硬化透明樹脂要使用硬化後也不容易黃變的清原 Craft 樹脂液。筆刀是拿來裁切假睫毛前使用，請務必要更換成新的刀片再行使用。樹脂的硬化處理使用的是美甲用的 36w 紫外線燈。軟橡皮擦是安裝假睫毛時，拿來固定位置使用。紙調色盤不只可以用於塗裝，也可以在裁切假睫毛、確認裁切下來的尺寸。還需要準備用來量測樹脂硬化時間的計時器、可以預先脂倒在上面備用的養生膠帶、還有用來保護塗裝完成零件的食品用保鮮膜。此外還要有牙籤蓋保護膠帶、鑷子組、將筆毛改造到只剩一根的自製面相筆等等道具。

Implanting eyelashes on the figure. The idea may come to everyone's mind once, but not many people will actuall it into action, learn it as a technique, and finish it as a work without feeling uncomfortable.This is a very patient pro but it is not impossible.These are false eyelashes for people that can be purchased at discount shops. The tip is around 0.4 ~ 1.2 mm to match the scale of the figure. First of all, tools necessary for hair transplantation are eyelashes. Chose the one with a sharp tip. The stereomicroscope has 20 times magnification. It is difficult to work the naked eye. The UV curable transparent resin is by Kiyohara Craft which hardly turns yellow after curing. H knives are used to cut the tip of the false eyelashes. New blades must be used. 36w UV light for nail is used for resin. The eraser is used as a guide when attaching eyelashes. Paper pallets are used not only for painting, but als cutting and checking the size. A timer for measuring the curing time of the resin, a curing tape are used for the liquid pallet and a plastic wrap for protecting painted parts. Other tools include toothpicks, masking tape, tweezers a hand-made brush with a single bristle.

取出裁切為 0.3～0.8mm 的假睫毛前端

Scoop the tip of the eyelashes that have been cut to 0.3 ~ 0.8 ㎜.

讓假睫毛的彎弧朝上放置

Place the eyelashes with curl upward.

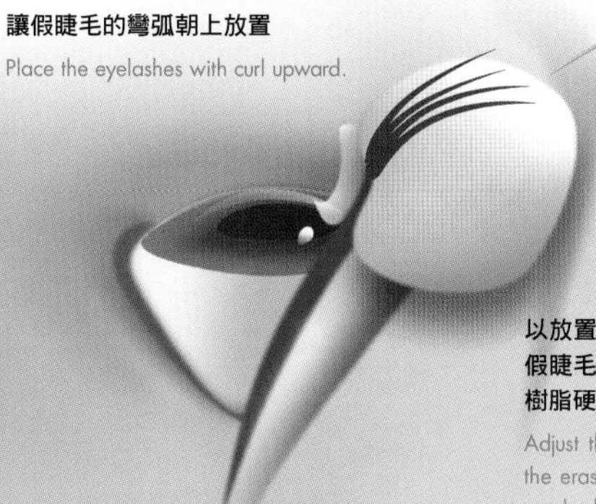

將筆毛裁切到只剩一根的自製面相筆。將筆尖沾上樹脂後待其硬化。

A original fine point brush made by cutting the tip of the brush into one hair. The tip of the brush is hardend with resin.

以放置在眼瞼上的軟橡皮擦為引導，整理假睫毛的角度。保持毛的位置不動，直到樹脂硬化為止。

Adjust the angle of the eyelashes according to the eraser guide placed on the eyelid. The hair can be held in place until the resin hardens.

接著作業結束後，將軟橡皮擦移開。由於燈源的熱度，有可能造成假睫毛變軟或是沾黏，所以要先等一會兒再將軟橡皮擦移開。如果無法順利取下的時候，請使用揉尖了的軟橡皮擦，將殘骸沾附去除。

Remove the eraser after bonding. The heat of the light softens the eraser and it may stick so leave the eraser for a while. If you can't remove it, make a sharp pointed eraser and stick it to remove the debris.

如果沒有使用軟橡皮擦來作為引導的話，假睫毛就會像右圖一樣散開。

Without the guide of the eraser, it will be loose as shown in the figure.

2 這是假睫毛的植毛概念圖。關於詳細的作業內容，請參考以下的步驟解說。安裝假睫毛的位置是在眼瞼的邊緣端面。藉由確定位置用的軟橡皮擦來調整假睫毛的角度，重點是要將每一根假睫毛都排列得整齊漂亮。田川 弘只要是 1/8 比例以上的人物模型，都會進行假睫毛植毛作業，但睫毛的長度與分量，還是要請各位實際進行植毛作業的時候自行判斷。此外，睫毛的角度也會因為眼睛半開向下看，或是睜大眼睛的狀態而有所不同。作業時一邊考量人物所處的場景環境，一邊調整睫毛的長度或角度是很重要的事情。說到這個作業的訣竅，有以下幾點。首先是使用加工成只剩下一根筆毛的自製面相筆，方便用來撈取裁切後的假睫毛；安裝假睫毛時，不要直接讓睫毛緊密排列，而是以間隔一根毛的距離開始作業；藉由暫時固定在眼瞼部分的軟橡皮擦，調整假睫毛的角度至排列整齊；以及使用 UV 硬化透明樹脂來固定假睫毛。大致以上這幾個重點。此外，因為這是相當精細的作業，所以還會使用到頭戴式放大鏡等道具。總之就是不要焦燥，提醒自己要慎重、仔細地作業。這的確不是一件容易的作業，但基本上只要實際經歷過這個植毛作業的過程，就會發現呈現出來的效果極大。作業過程中殘留在指尖的觸感，就是我們將這門技術已經納為己用的證明。

Conceptual diagram of eyelash hair transplantation. Please refer to the following process for details. This section introduces, with easy-to-understand illustrations, which parts of the eyelid should be coated with UV-curable transparent resin and the cut fake eyelashes should be attached. The key is to control the angle of the eyelashes with an eraser for positioning the end face of the eyelids so that the eyelashes are aligned neatly. Tagawa implants eyelashes on a scale of 1/8 or larger, but the length and volume of the eyelashes have to be determined on their own after they are actually implanted. The angle of the eyelashes also varies depending on whether they are slightly lowered with downward-looking eyes or the pupils are clearly opened.In such a situation, it is important to work while considering the length and angle of the eyelashes. The key points are to use a face brush with a single tip to scoop up the cut false eyelashes; to attach the eyelashes, do not i mm ediately line them up closely but leave an interval of about one; align the angle of the eyelashes with an eraser temporarily fixed to the eyelid; and to fix the eyelashes, use a UV curable transparent resin. Also, since it is a delicate work, it is recommended to use tools such as a head mounted magnifier , and do the work carefully and carefully without rushing. It's not an easy task, but basically, once you actually experience the hair transplantation process, the effect is tremendous. The feeling of the fingertips can be learned as a technique.

3 將已經塗裝完成的人物模型本體包覆起來保護。常常會有不知不覺將零件在手裡握得太緊，或是碰撞到哪個地方的可能性，所以這個步驟是一定不能省略跳過。將零件整個用食品保鮮膜纏繞起來。對於形狀不固定，而且需要保護的面積很大的人物模型來說，保鮮膜是非常方便用來包覆整體的道具。頭部因為無法纏繞保鮮膜的關係，所以貼上保護膠帶代替。

4 將假睫毛拆散。使用筆刀慎重地進行裁切作業。有時候裁切下來的假睫毛會因為靜電的關係，沾黏在刀片上，此時可以用牙籤來將假睫毛取下。田川 弘自從改在紙調色盤上進行這項作業之後的效果極佳，100% 再也沒有弄丟過任何一根裁切下來的假睫毛。

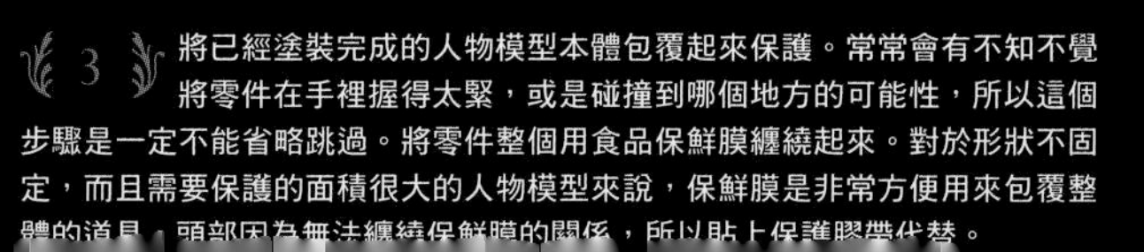

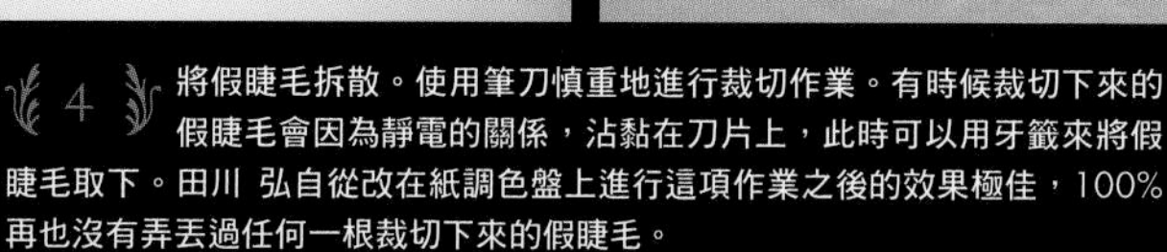

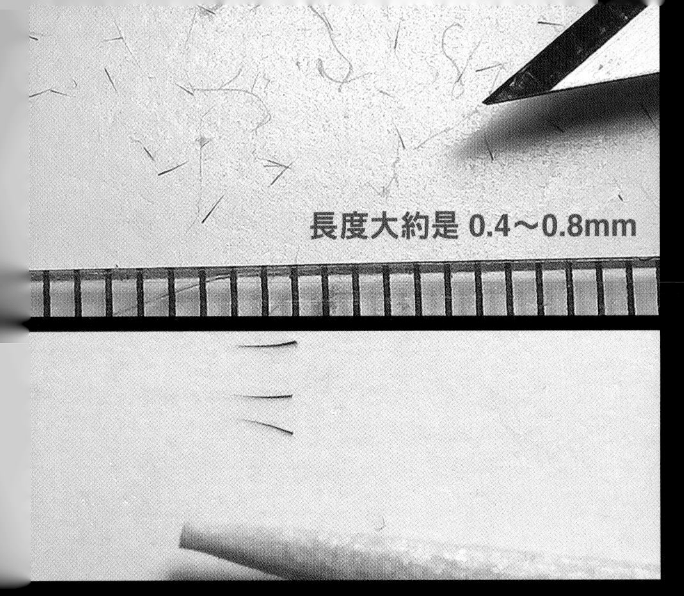

5 放上直尺來作為比較。原則上 1/8 比例的話是將長度裁切成 0.4～0.8mm。裁切下 4～5 根後，將假睫毛拿到預定要接著固定的位置，確認是否符合自己想要呈現的氣氛。一開始尚未熟練時不好控制尺寸，可以裁切得稍微長一些，但是完成時有可能會因此感覺到有哪裡不對勁。這部分的拿捏就需要依靠經驗了。

The tip length is 0.4 ~ 0.8 mm for the 1/8 scale figure. After cutting 4 ~ 5 eyelashes, take them to the place where you want to attach them and check the appearance. It's a good idea to cut it longer for the first few times, but finbal result may feel a little uncomfortable. Experience counts in this area.

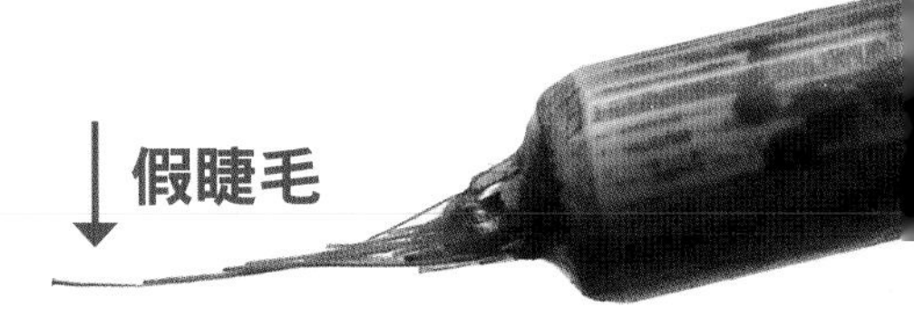

6 如果用鑷子去夾取裁切下來的假睫毛的話，形狀會因此扭曲變形，萬萬不可。這裡要使用加工成只剩下一根筆毛的面相筆，先將筆毛用 UV 硬化樹脂稍微沾濕，再以此將假睫毛撈起來，移至想要的位置。如果這個狀態下的假睫毛翹曲角度沒問題，移至接著位置觀察也沒有違和感的話，再繼續裁切相同長度的假睫毛 30 根左右。

If you pick up the tip of the cut eyelashes with tweezers, the shape will be distorted. A single-tipped fine brush is moistened slightly with UV curing resin to scoop up and move the tips of the eyelashes. Check the curvature of the eyelashes at this state, When you are satisfied with it, make about 30 more eyelashes with the same length.

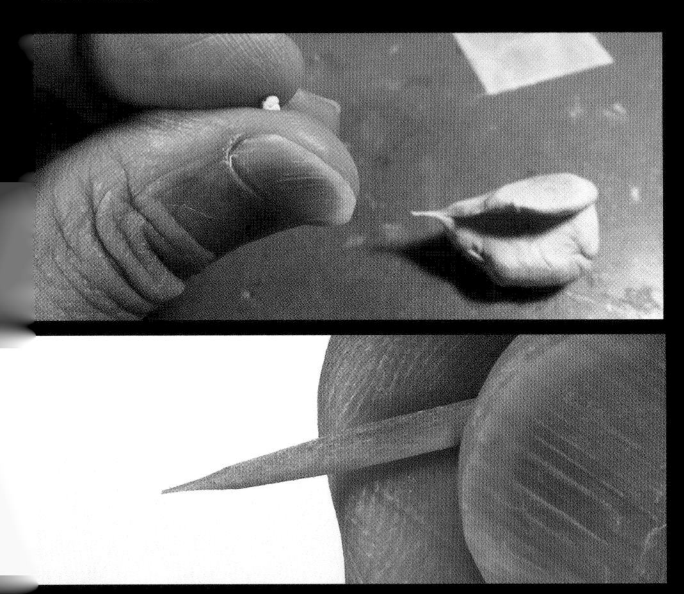

7 使用軟橡皮擦製作假睫毛接著時的引導用底座。這可以用來幫助讓假睫毛的安裝角度整齊美觀，並在假睫毛接著使用的 UV 樹脂硬化前，維持假睫毛的位置固定不動。除此之外，使用面相筆將假睫毛移至接著位置時，軟橡皮擦本身的黏著力也可以方便將假睫毛自筆尖取下，算是附加的效

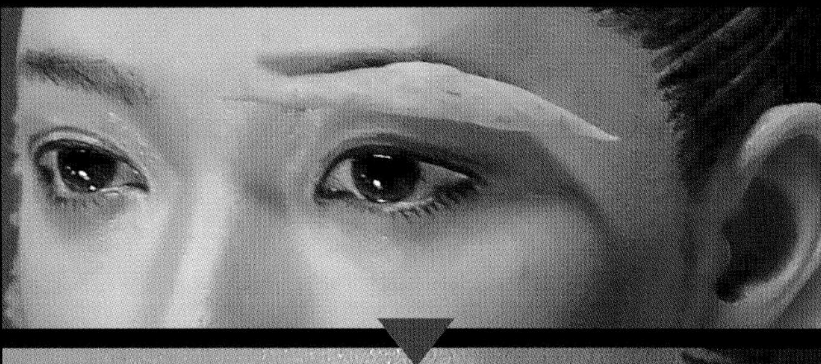

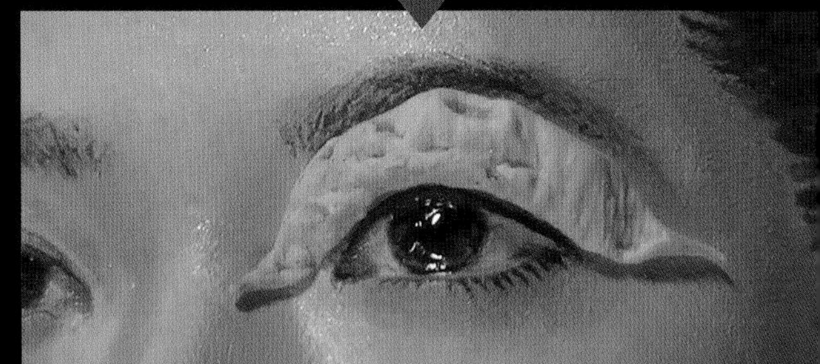

8 將軟橡皮擦捏塑成照片上的這個形狀，然後輕輕地貼附在想要的位置上。接下來使用前端裁切成平頭的牙籤，將軟橡皮擦按壓移動至任何自己想要的位置。此時，如果作業前先將牙籤的前端沾濕的話，比較不容易沾黏上軟橡皮擦。關於軟橡皮擦的位置及厚度，請參考第 29 頁插畫的切面

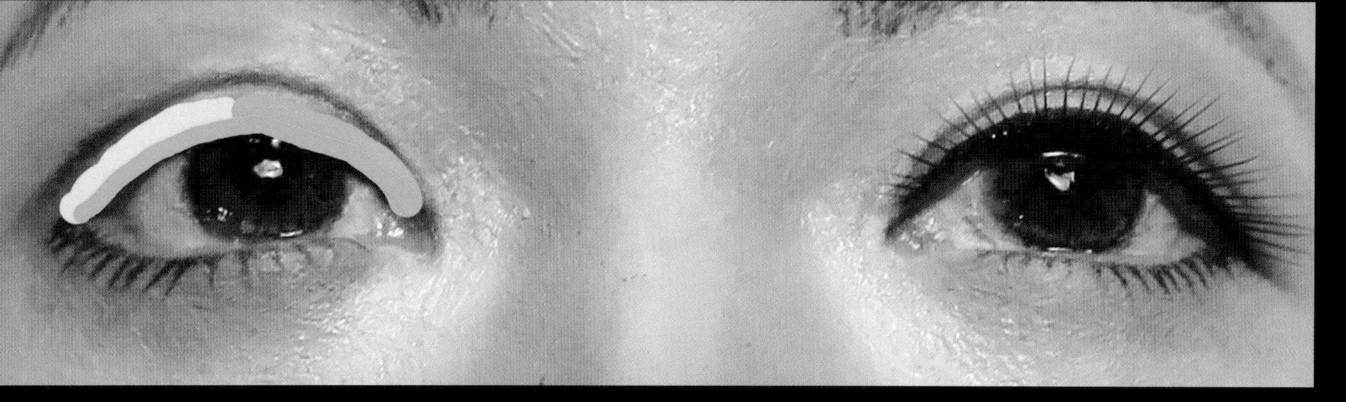

9 雖然植毛作業可以一口氣完成，但是因為集中力不容易持續這麼久的關係，分成 3 道步驟對於精神方面的健康會比較好。首先在照片上黃色線條標示的區域，以只有一根筆毛的面相筆塗上 UV 硬化樹脂。配置假睫毛的作業重點是彼此要保持一根毛的間隔。如果排列緊密太過勉強的話，有可能排列會產生歪斜，或者是兩根睫毛沾黏成一根毛，千萬要注意。黃色部分植上睫毛後，以 UV 燈照射 6 分鐘，使樹脂硬化。接下來是在橙色線條的位置塗上 UV 樹脂，同樣保持一根毛的間隔距離進行植毛，並使樹脂硬化。最後要在綠色線條的部分塗布 UV 樹脂，將先前植毛作業保留下來的間隔也植上假睫毛，然後照射 UV 燈。硬化後再一次將整體塗上 UV 樹脂，以 UV 燈照射 12 分鐘後就告完成了。

Eyelashes can be attached all at once, but it is also good to divide the work into 3 steps because patience does not last long. First, a UV curing resin is applied to the part shown by yellow lines in the image with a brush. The point is to leave enough intervals between the eyelashes. If you force the eyelashes to stick, the adjacent eyelashes will interfere and distort, or the two eyelashes will stick to one, so be careful. After placing the eyelashes on the yellow part, the resin is cured by UV light for 6 minutes. Next, the UV resin is applied to the orange line part, and similarly cured. Finally, UV resin is applied to the green line part, the eyelashes are placed in the gaps between already attached eyelashes. After all the steps, the entire surface is coated with UV resin and irradiated with UV light for 12 minutes to complete.

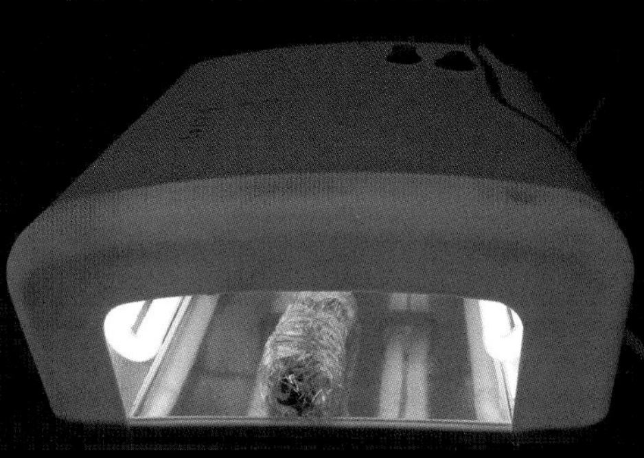

10 當最後的 UV 燈照射作業完成，確認假睫毛已經固定住後，將引導底座的軟橡皮擦移除。如果因為 UV 燈的熱度造成軟橡皮擦軟化不容易一次就完全移除乾淨的話，可以將軟橡皮擦的前端捏尖，按壓在殘留下來的碎屑上即可方便去除。如果使用牙籤或是鑷子這類堅硬的工具，有可能會傷害到已經完成的塗裝面，萬萬不可。耐住性子不要焦急，花時間慎重地作業才是距離完成最短的捷徑。一側的植毛作業完成後，接著在另一側進行相同的作業。熟練之後，單側的作業時間大約是 30～40 分鐘左右即可完成。只要學習到這個植毛技術，相信一定能讓人物模型的完成度提升一個層次。

When the last UV light irradiation is completed and the fixation is confirmed, the eraser of the guide is removed. The eraser is softened by the heat of the UV light and cannot be removed cleanly at a time. In that is the case, pressed the sharpened eraser against the remains. Do not use toothpicks or tweezers as they may damage the painted surface. In the end, the shortest way is to work carefully and slowly. When one side is completed, the same operation is applied to the other side. Once you get used to it, it takes about 30 to 40 minutes to complete one side.

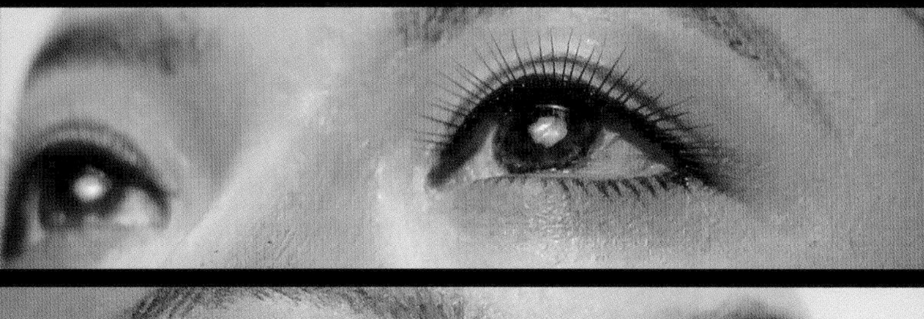

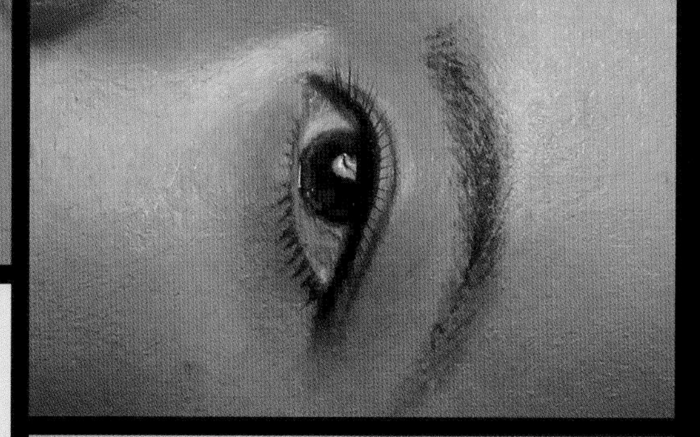

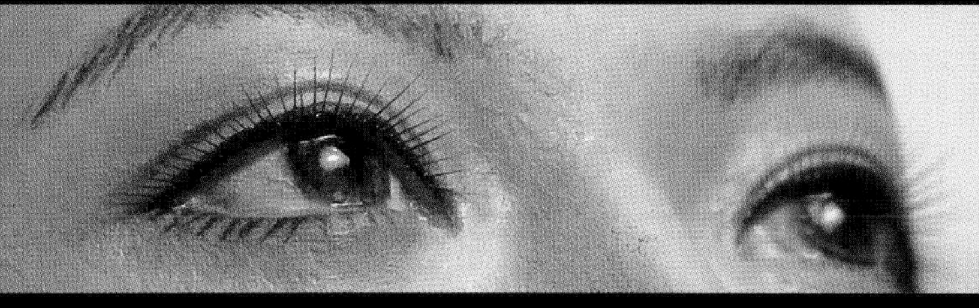

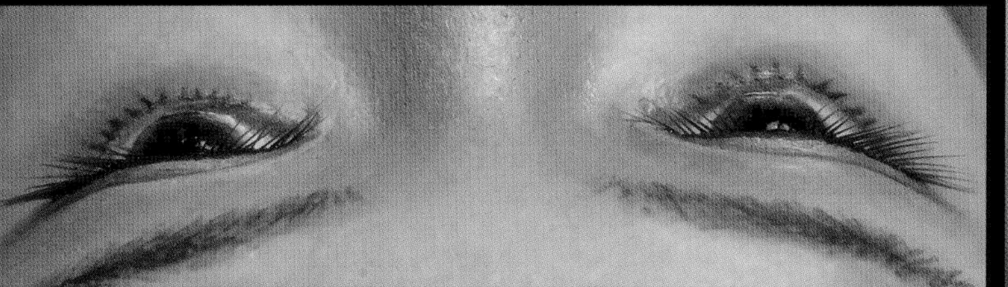

肌膚的部分使用 Finisher's 的 Multi Primer 複合底漆補土進行底層處理，然後再以 Gaianotes Surface Evo 的粉紅色均勻塗在整個零件上。接下來的作業要一邊意識到形狀的凹凸，一邊從斜上方 45 度角的位置將 Surface Evo 的 Fresh 塗料以空氣噴筆進行塗裝。噴塗作業時要刻意保留細微凹陷處的粉紅色。再來就是戴上頭戴式放大鏡，確認外觀有無異物、灰塵以及扭曲的凹凸不平。如果找到的話，使用研磨砂紙將其去除，或是以瞬間彩色補土來進行修正。

For the skin, the multi-primer of Finisher's is applied to the base, and the pink of Surfacer Evo of Gaianote is sprayed evenly over the entire body. Next, while being conscious of the unevenness, the fresh color surface evo is sprayed with air brush from upper oblique angle of about 45 degrees to leave pink on the recessed place on the body. Finally, use the head magnifier glasses to check for dust and uneven surface. If they are found, they are removed with sandpaper and corrected with instant color putty.

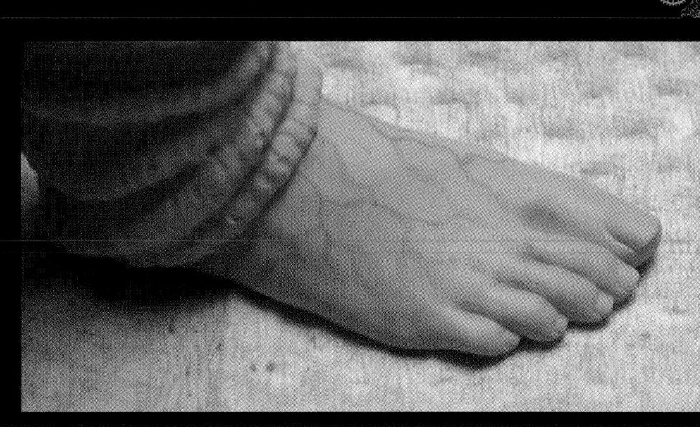

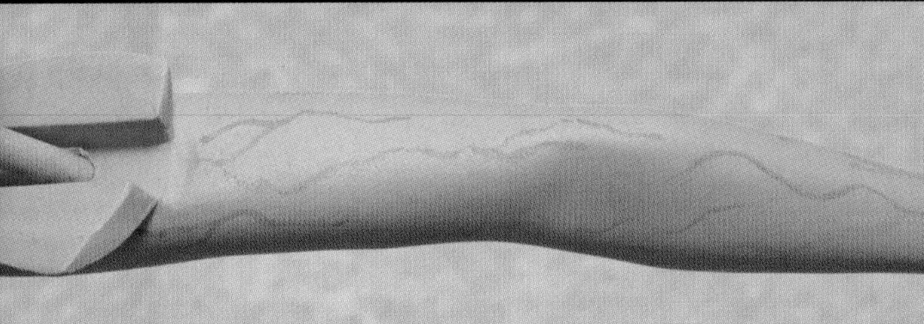

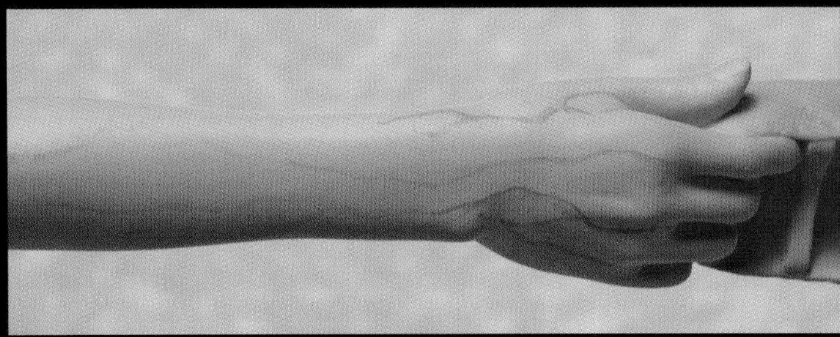

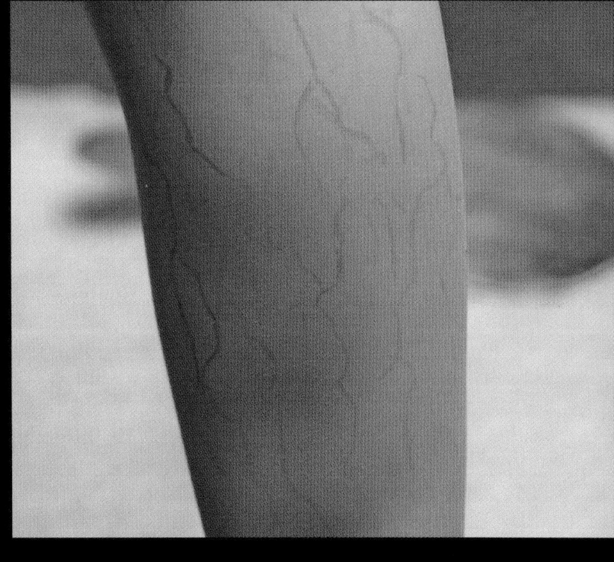

1 肌膚塗裝所使用的是油畫顏料。由 KUSAKABE 公司生產的鈦白色、淺鎘紅色、檸檬淺鎘黃色、鎘黃色、鎘綠色以及群青色。除此之外也會使用到 Holbein 好賓公司的淺洋紅色、水手藍色等等油畫顏料。調色盤上預先調好接近人物模型上塗裝顏色的肌膚色以及稍微較深的肌膚色備用。首先要在大腿、上臂、手掌、腳掌、頸部這些皮膚較薄的部分描繪出血管的紋路。血管的顏色是將先前混色準備好的肌膚色，加上鎘綠色與群青色進一步混色調合而成。使用的畫筆是筆毛較長的日本畫面相筆，這是因為面相筆一次可以沾附顏料的量較多，有可以藉由筆壓的強弱來自由控制線條粗細的優點。下筆時，藉由指尖有意識的「晃動」，可以表現出更加寫實的血管扭曲形狀。如果不慎將血管描繪得過粗或是畫錯位置時，可以用指腹按壓在想要修改的線條上，使其模糊變淡或是整個抹去。接下來皮膚的顏色還會有 1～2 個階段的疊加上色，所以只要是處理到不明顯即可。反過來說保留一些這樣的線條，後面也許還會形成意料之外的效果，因此這裡不需要太過神經質也無妨。以好的方向來說，隨興的「差不多即可」也是有必要的。血管描繪完成後，放入烘碗機內 1～2 天使顏料乾燥。

Oil paint is used to paint the skin. Colors are KUSAKABE's titanium white, cadmium red light, cadmium red deep, cadmium yellow lemon, cadmium yellow deep and peach black and in addition, Horbein's Light-Magenta, Marine Blue, etc. A color similar to the color painted on the figure on the palette and a slightly darker skin color are mixed and prepared. First, draw blood vessels on thin skin areas, such as the thighs and upper arms, the First, draw blood vessels on thin areas of skin, such as the thighs and upper arms, the insteps, and the back of the neck. The Japanese long fine point brush has the advantage of being able to draw lines freely from thick to thin with different pressure. The conscious "shake" of the fingertips can represent the distortion of the blood vessels.If you feel that you have made a mistake, use your fingers to blur or erase it. After that, the skin color will be laminated in 1 ~ 2 layers, so it is enough to make it inconspicuous.

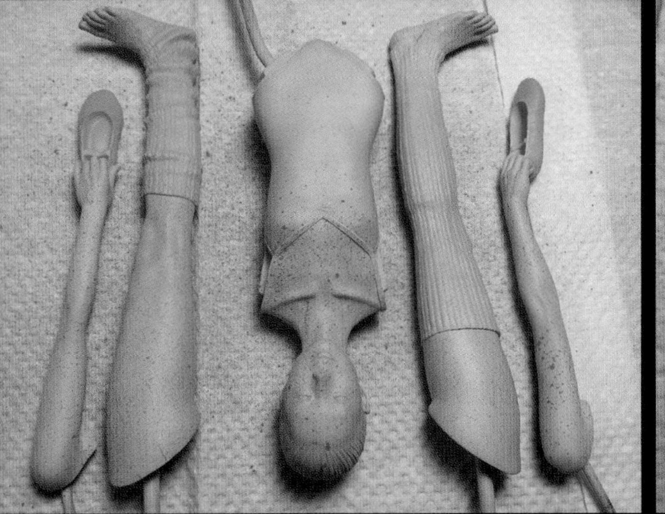

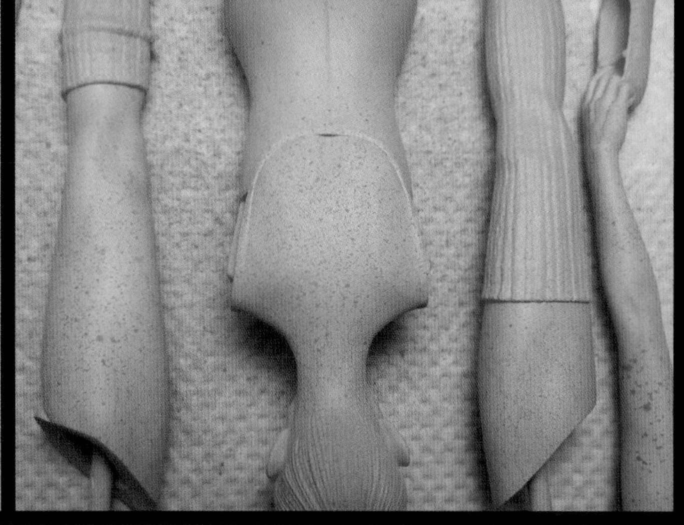

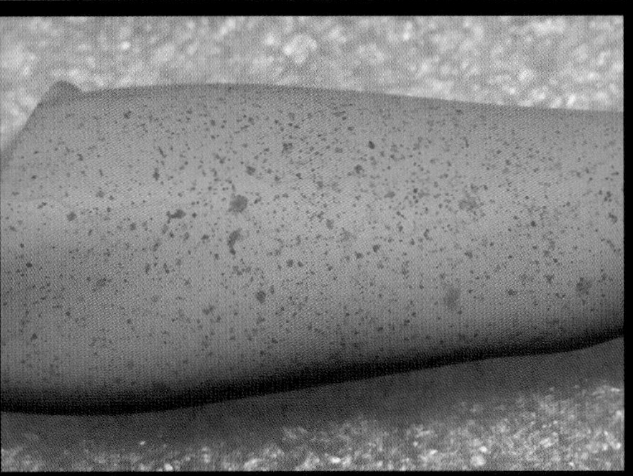

2 田川 弘的人物模型塗裝，除了在底層塗裝之後的血管描繪表現以外，另一項特徵就是在表面施加沒有特定形狀的「斑點」表現。先準備好將鎘紅色與淺洋紅色油畫顏料以溶劑油稀釋後的顏料，此時的黏度不要像水一樣稀薄，而是要稍微保留一些帶有黏性的感覺。接下來以較硬的筆刷前端沾附顏料後，用指尖像是翻書過頁一般撥動筆刷，利用刷毛本身的彈性，將顏料噴濺在零件上。當零件整體都以這個方法完成斑點塗裝後，放入烘碗機內烘乾 2～3 天，直到完全乾燥為止。之所以相較於其他底層塗裝耗費更長時間進行硬化處理，理由是為了要確保在之後的上層塗裝疊色作業時，絕對不會讓這裡的塗裝顏色溶化。人物模型的塗裝，尤其是以女性為創作主題的動漫角色人物模型，一般都會避免在肌膚上進行和暗沉效果有關的表現，但人類的肌膚如果在沒有施加粉底修飾的狀態下，一般看起來往往「並非那麼乾淨漂亮」，這和人種不同、膚色不同完全沒有關係。請各位試著仔細觀察一下自己的手臂，除了可以看到透出皮膚的血管顏色之外，也應該可以理解到皮膚的顏色絕對不是完全一致沒有差異的。這裡所做的「斑點」處理，正是將這個視覺資訊具現化之後的狀態。此外，所選用的顏色也能讓觀看者聯想到皮膚底下的肌肉組織的顏色，形成更具備真實感的質感呈現。

As a feature of figure painting of Tagawa, an amorphous expression of "stain" is used in addition to blood vessel expression. A mixture cadmium red and light magenta dissolved in petrol is prepared, and the paint viscosity in this case is slightly sticky. Soak the paint in the hard tip brush, and use the elasticity of the brush to splash the color like if you are flipping the book with your fingertips.After applying this stain to the whole body, let it completely harden in a dryer for 2-3 days. The reason it takes more time than other base coatings is that the color of the paint doesn't melt when you paint over it.n the case of painting figures, especially female character figures, it is often avoided to express dullness on the skin in general, but without makeup foundation, it is common for human skin to be "not so pretty" which is not related to the color of the skin by race.If you take a close look at your arm, you'll see that it's not a uniform color, This "stain" embodies such visual information and

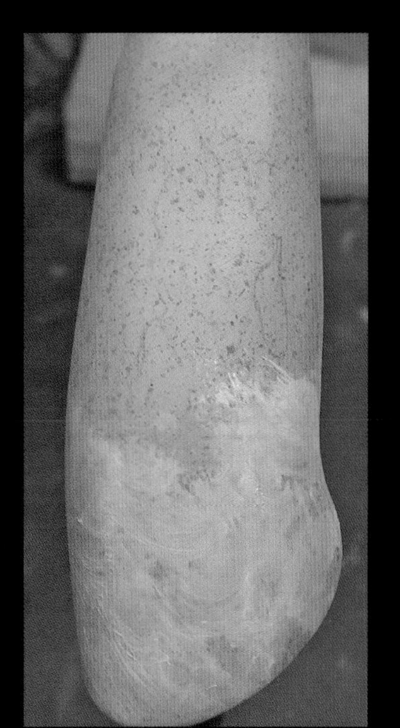

3 以油畫顏料開始進行皮膚的第一層上色。使用一開始預先調製的肌膚色作為基本色，加入大量的白色後調合成這裡需要的顏色。首先是為大腿根部到大腿中段以及手臂的根部到手臂中段、手肘的內側、胸部、側腹部等等，日常生活中不會受到陽光直射的部位上色。製作範例中雖然使用的不是肌膚色，而是直接使用白色顏料，但顏料是調合成相當稀薄的狀態，所以會像照片上一樣，底層的基底色穿透至上層，修飾後的效果看起來並不像是直接使用白色顏料。當然，我們需要視塗裝的對象、創作主題以及故事背景來臨機應變調整肌膚的色調。

The first layer of skin with oil paint. First,prepare a mixture of skin colors with a lot of white added. From the groin area to the middle of the thigh, from the shoulder joint to the middle of the upper arm, the other side of the elbow, chest, side, and other areas that don't receive direct sunlight on a daily basis. In the example, the white is used instead of the skin color, but the base color is transparent as shown in the image because the density of the oil color is very thin. Naturally, the color of the skin changes according to the object, motif and story to be painted.

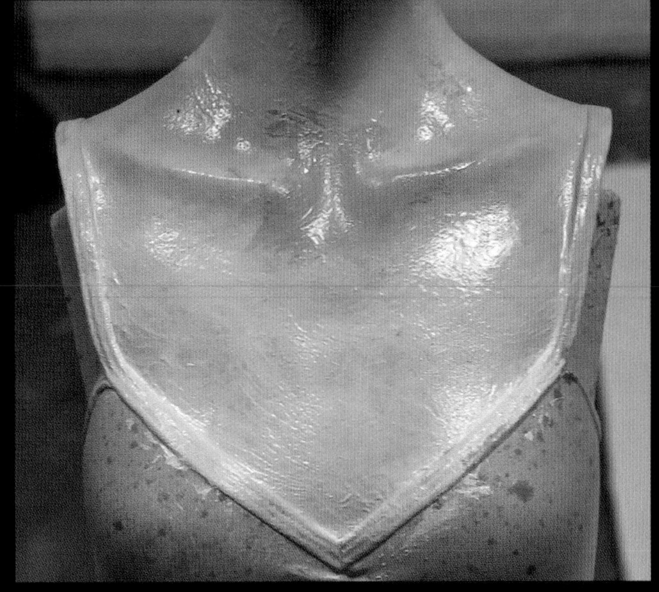

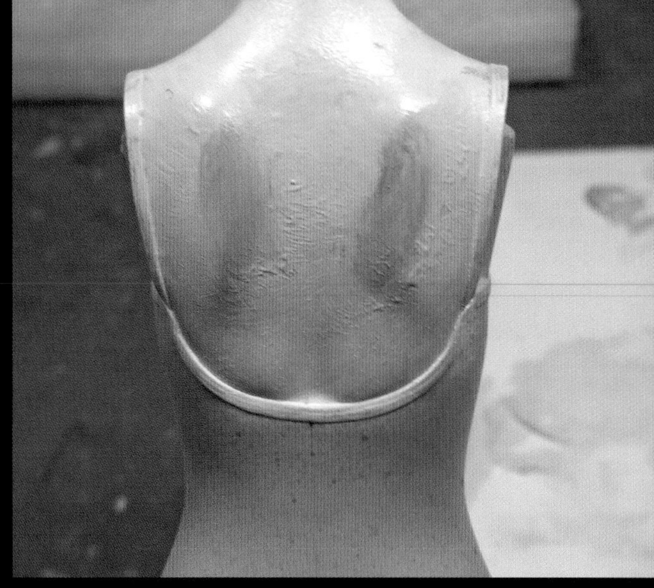

4 在基本的肌膚色加入鎘紅色與淺洋紅色進行混色後，為鎖骨的凸出部位上色。田川 弘並非如一般的作法在「零件凹陷形狀的部位」使用陰影色上色，而是以在皮膚加上紅色來襯托的方式來表現。人物模型的塗裝為了要增加視覺上強弱對比的效果，經常會在形成陰影的部位以較暗的色彩呈現，但這種作法對於田川 弘來說，因為最終還是會架設燈光來拍攝成平面的作品，所以從基本的方法論就已經採行與其他人不同的方向。

A shade of cadmium red and light magenta is placed between the collarbones. Note that not all "a concave part" are shaded. In painting a figure, there are some cases where shades are added in order to make the figure look sharp, but this is not necessary from Tagawa's method of setting up lighting and taking pictures for final work.

5 與鎖骨相同，在肩胛骨的凸出部位也加上偏紅的色彩後，使用將筆尖切平的自製戳染專用畫筆，不沾任何顏料，以接近垂直的角度，用筆尖平面在塗裝面上以敲打戳染的方式使顏色與周圍暈染融合。決定好身體的中心線，先由右側朝正面塗裝→再由左側朝正面塗裝，左右對稱塗色，並且由色調較淡的部分開始下筆。敲打戳染的過程中，顏色會自然地留在筆尖上，使塗裝面呈現出整體統一的感覺。

In the same way, the shadow color is placed in the hollow part of the scapula. When you have finished placing the shadow color, use the tip of self made striking brush to put the basic color at an angle almost perpendicular to the surface. The central line of the body is determined, and progress symmetrically from the right chest to the left chest and from the lightly colored area.

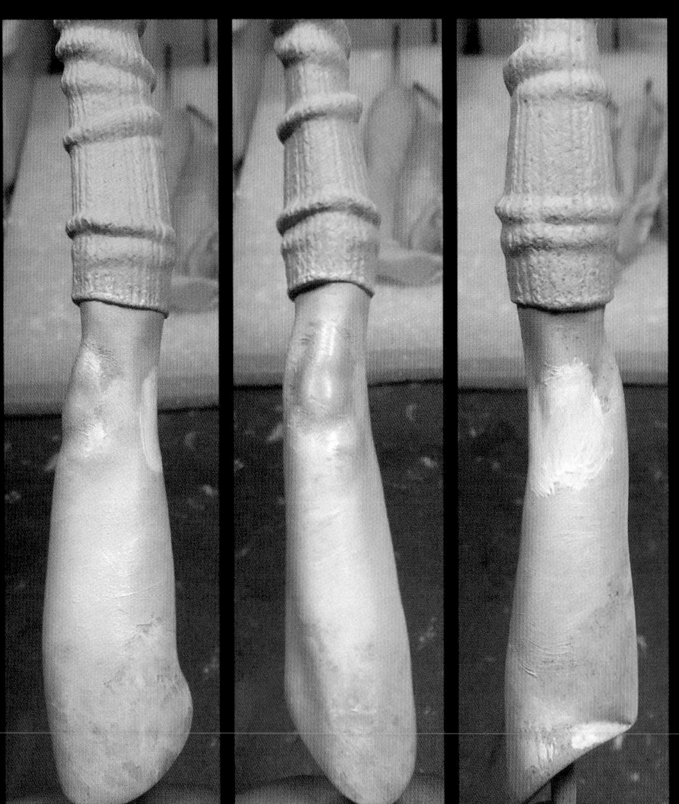

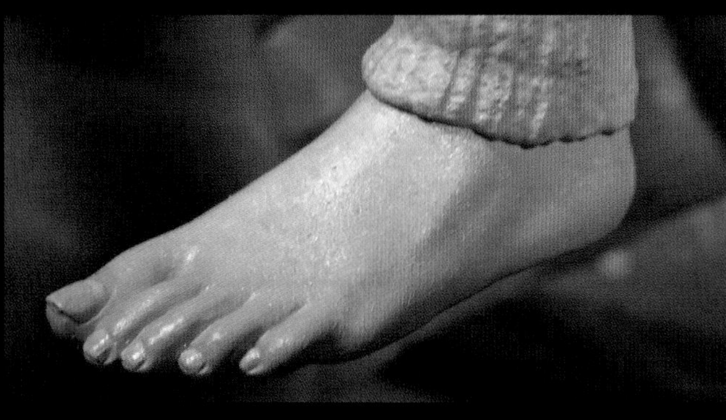

6 在基本色的肌膚色中極少量混入色相環中屬於相反色的綠色，讓彩度降低，然後再稍微摻入紅色來調製成色調略深一些的肌膚色顏料。將這個顏色在手肘及膝蓋部位上色。小腿肚是以淺洋紅色上色；手指及腳趾、肚臍是以基本的肌膚色加上少量鎘紅色與淺洋紅色混色後的顏料上色。手指前端這類面積較狹窄或是曲面較小的部分，以色調略深的顏色上色可以強調出陰暗的感覺，呈現出更加自然的血液流通部分的氣氛。這樣的塗裝方式與其說是寫實，更不如說是繪畫的表現手法。所謂的變形強調的表現，並非只是單純的陰影色演出，而是藉由四肢的末端來表現出人物的躍動感及生命力，並且讓所有觀看作品的人都能夠充分感受到這股活力。

The basic skin color is mixed with a very small amount of green, which is the opposite color in terms of hue, to reduce the saturation, and then a darker skin color is prepared by mixing a little red. Put it on the elbow or knee.A light magenta is placed on the calf. On the fingers, toes and navel, a mixture of cadmium red, light magenta and the basic skin color is used.In areas such as fingertips where the area is tiny or the curvature is small, the shade is emphasized by making the color a little darker, resulting in a more natural look. It is more picturesque than realistic, and it is a so-called deformed expression, but it is not a mere shadow color production, and it contains the expression of lively feeling and vitality from the end of limbs, and it is sufficiently transmitted to people who see the work.

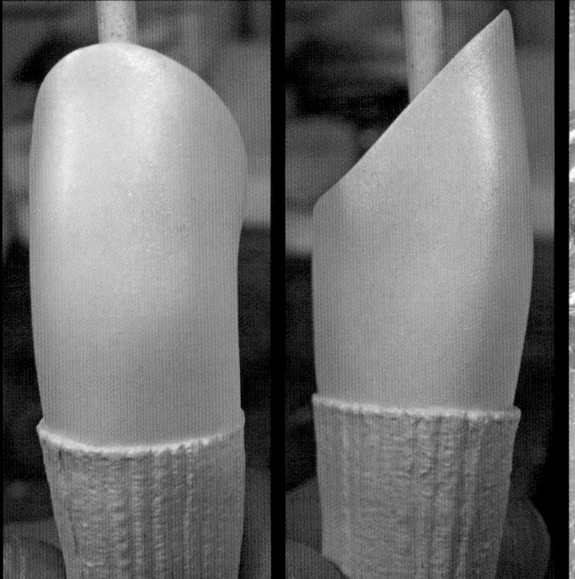

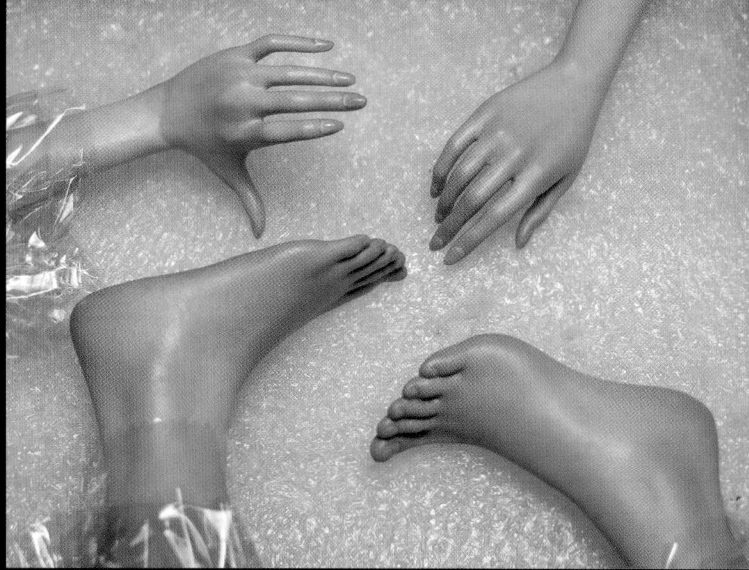

7 接下來要進行手腳指尖的修飾步驟。各個關節以及指甲的生長邊緣、指甲上的半月、還有手掌要以線條描繪的方式來呈現。在手指之間及腋下的凹陷部位以及服裝與肌膚的交界線要加上暗色調。指甲使用透明琺瑯漆先塗上一層保護層，放入烘碗機乾燥一天，然後再使用購買於 3 年前黏性已經變得較高的透明琺瑯漆重新塗上一層保護層後就完成了。最近偶爾也會改用 UV 硬化透明樹脂。

Finishing of fingertips and toes. Draw lines on each joint, nail and the palm of the hand. Put the shadow in the dents between the fingers, the sides and around the edges of the costume and skin. The nails are coated with clear enamel paint and dried for a day, and then over coated with another slightly more viscous enamel clear. UV curable transparent resin is used sometime for final over coat.

8 在開始進入第 2 層油畫顏料上色前，先要以布製的研磨砂布輕輕地磨擦較大面積的塗裝面 1～2 次，進行整平。身體使用#1000、臉部使用#1000 與#1500 砂布，細部細節的部分則只需要處理覺得不滿意的地方即可。第 2 層上色是以更偏向白色的肌膚色來薄薄地疊上一層顏色。以 Holbein 好賓的溶劑油將顏料稀釋成接近水的液狀，然後以再向外擴散般地感覺進行塗布。作業時要去意識到這一層顏色將會成為肌膚的上層部分，彷彿要用這個顏色將整個零件都包覆起來般的筆觸進行運筆。最後還要以筆毛較軟的自製戳染筆的筆尖來輕輕敲擊塗裝面。此時的訣竅是以要將塗料以輕輕推開的感覺，一邊溫柔地敲打。

Before working on the 2nd layer of oil painting, lightly rub the large painted surface with cloth file 1 ~ 2 times. # 1000 for the body, # 1000 and # 1500 for the face, and other details where you are concerned. The second layer is painted with a highly diluted lighter whitish skin color. As you realize that this very thin layer will become the upper layer of your skin, spread it out with a brush as if wrapping it all in this color. In the end, you hit it with a soft tapping brush. The point is to hit it gently so the brush will stretch the paint.

油畫顏料的關鍵在於乾燥

The Dryig is the key of the Oil paint.

底層塗裝雖然使用的是硝基塗料，不過肌膚等處的正式塗裝使用的是稱為油彩的油畫顏料。以模型專用塗料來說，比較接近這個顏料特性的是琺瑯漆。但是油畫顏料的良好發色以及色數選擇之多，還是沒有其他塗料能出其右。使用上的瓶頸點在於需要較長的硬化（書中有時也稱為乾燥）時間，但田川 弘使用市面上販售的烘碗機來作為因應方式。不論冬、夏季節，溫度都設定在 36～38 度左右。幾乎任何一種塗膜經過 1～2 天的烘乾之後，都可以乾燥到用手直接觸摸也不掉色的程度。就是受惠於這個能夠加速乾燥的道具，方能實現以油畫顏料進行塗裝的技法。

Lacquer colors are used for the base coating, and oil paint is used for the main layer such as the skin. Enamel color is close to this type of paint for modeling, but number of colors and vividness are not comparable to the oil paint. The length of the hardening (drying) time is longer, but Tagawa copes with this by using a dish dryer. By setting the temperature to about 36 ~ 38 degrees Celsius regardless of the season, most coatings can be cured within 1-2 days without problems. With this speed, the painting with oil paint can be done quickly.

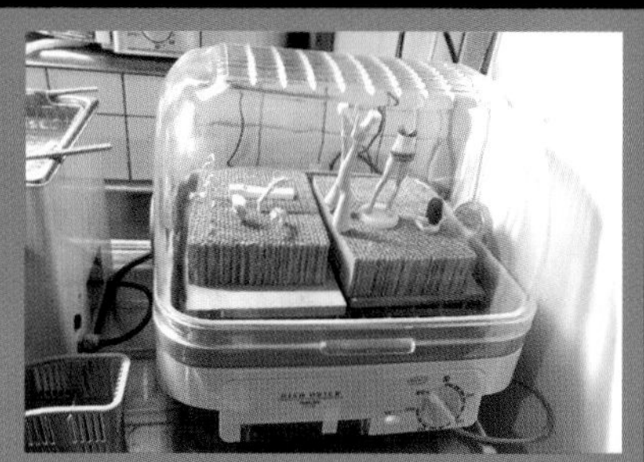

除了斑點、血管這類位於塗裝面底層的細節演出之外，在皮膚表面描繪出「黑痣」也是田川作品的一大特徵。田川 弘 20 年前還只是在身上 1～2 處加上黑痣而已，但最近 5 年來已經進化成在全身各處都加上了無數的黑痣。這樣一方面可以用來遮掩在快要完成之前才發現到的微小氣泡或是傷痕；另一方面，斑點、皺紋、黑痣這些乍看之下讓人覺得屬於負面的缺點項目，實際上卻能夠形成對照，成為襯托出整體美麗的要素。

Tagawa's works feature "mole" painted on the surface of the skin in addition to the stage effects applied to under layers of painted surfaces such as stains and blood vessels. He started put innumerable number of moles on the whole body of the figures. It also helps hide tiny bubbles and scars that were discovered just before completion, but those spots, wrinkles, and moles, enhance the overall beauty.

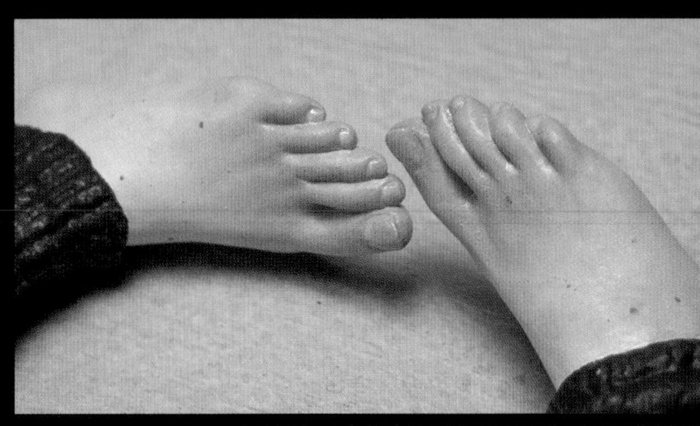

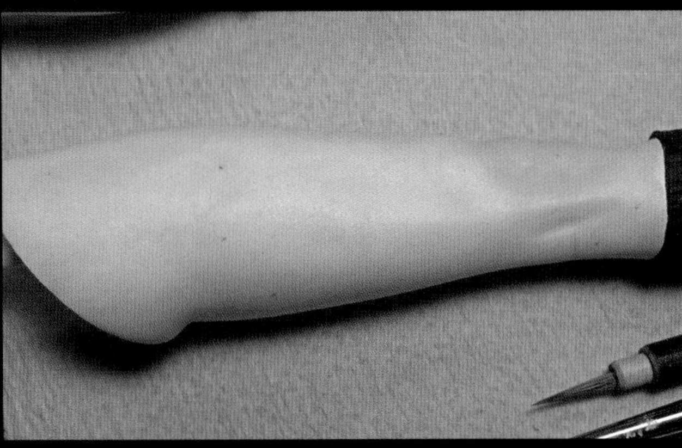

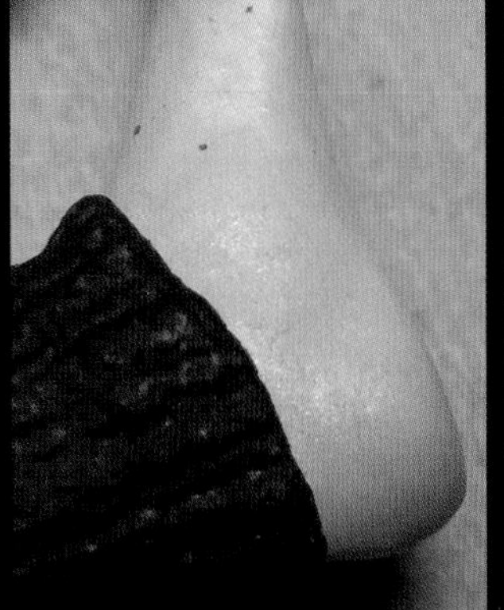

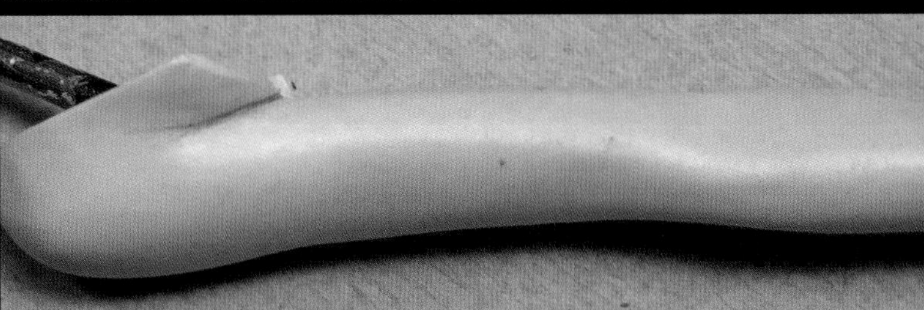

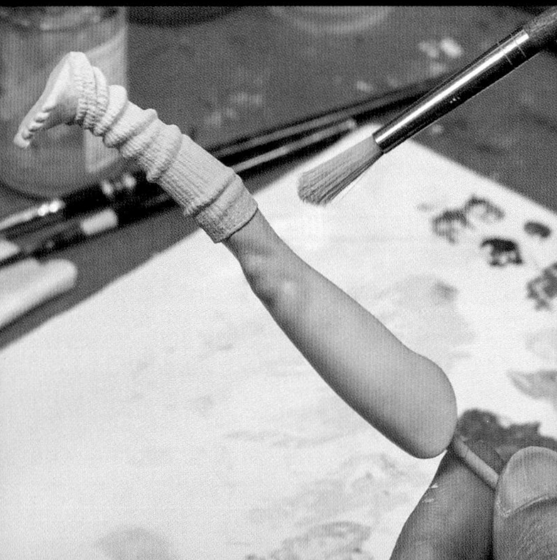

1 手肘與膝蓋這些部分的皮膚通常都處於收縮的狀態，所以外觀看起來不會那麼緊繃。也因此與身體其他部分相較之下，屬於比較感覺不到光澤的部分。反過來說，像是手指的第三關節或是手腕關節的凸出形狀、腳踝部分等等，就是要以紅色調來襯托出外觀呈現緊繃狀態的部位。此外，裸露在衣服外面的肌膚部分，也會因為部位的不同，而有除了色調之外，還有皮膚光澤感、色差不均感等等需要控制調整，藉以表現出更讓人感受到生命力的「栩栩如生」的表現。還有，平常不容易曬到太陽的部分，會以調成稀薄狀態的白色油畫顏料直接塗色上去。藉由自製筆刷的戳染技法，可以讓白色與周圍的肌膚色暈染融合在一起，而且底層的基底色也會透出浮現於外表，所以在完成前的最後修飾階段，幾乎讓人感覺不到曾經使用過白色顏料。

Areas such as the elbows and knees are usually not firm because the skin is contracted. Therefore, compared to other parts, this part does not have a luster. On the other hand, spots like the third joint of the fingers, the wrist joint and the ankle are reddish and tensed. Even if it is an exposed part of the skin, by controlling not only the color but also the gloss and unevenness of the color depending on the part, it becomes an expression that feels more "alive". In addition, he usually paints the parts hidden from the sun with thin white oil paint. Because of the tapping technique with a brush, it mixes with the skin color around it, and the base color of the base can be seen through, so that it doesn't look like just white paint at all.

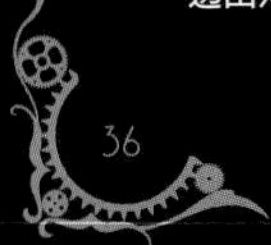

頭髮的塗裝 Hair

有很多擬真類的人物模型，在形塑零件外觀造型的時候，就已經會將頭髮的流動感仔細的呈現出來。除了髮色的表現之外，對於擬真類的人物模型來說，如何能夠重現髮際線的質感，想必也是佔有相當重要的比重。配合外觀造型將髮際線的流動感描繪出來，藉由髮際線的位置控制臉部面積的比例，可以讓完成後的表情產生相當大的變化。這是與化妝不同的另一個可以影響表情的重要項目。

In the case of realistic figures, it is not rare to see even the finer flow of hair molded into parts. Not only the expression of the hair color but also how to reproduce the texture of the hairline will surely be an important factor for realistic figure. By drawing the flow of hair at the hairline along the mold and controlling the area ratio of the face part, the finished expression can be greatly changed. It is the point of applying a different expression from makeup.

1 這個套件將頭髮外形的凹凸起伏都製作得相當漂亮。我們要利用凸出部分的形狀將髮際線描繪出來。由於原始的套件讓人感覺額頭有些過於寬廣，因此可以藉由直接在凸出形狀的延長位置上色，來改變整體印象。不論是擬真類或是動漫角色人物類的人物模型，瀏海的髮際線位置都會帶來整體印象的變化，對於最後完成的狀態產生極大影響，所以髮際線是必須要特別注意的重點事項。馬尾造型的頭髮因為受到牽扯的關係，髮際線的皮膚也會呈現緊繃的狀態。此外，因為皮膚下面有頭髮毛根的關係，在外觀上呈現出來的色調會與肌膚不同。由於太陽穴的部分沒有將頭髮的外觀造型製作出來，所以在這裡描繪的髮量也會影響人物的整體氣氛。畢竟頭髮本來就是構成人物表情的重要部位。

In this kit, even the unevenness of the hair is beautifully molded. This convex part is used to draw the hairline part of the hair.Since he felt that the forehead was a little wide with the kit, he changed the overall impression by placing the color on the extension of the convex mold. It is important to note that the impression changes depending on the position of the hairline of the bangs, whether it is real type or character type, greatly affect the finish. The skin of the hairline is also swollen because it is pulled by the hair. It also has hair roots under the skin, so it looks different. The temple part is not molded, so the amount of hair drawn here also changes the atmosphere. Hair is also an important part of facial expression.

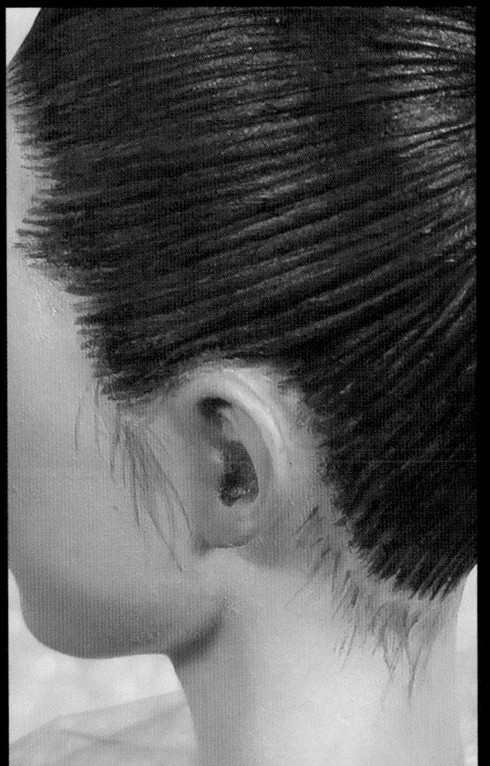

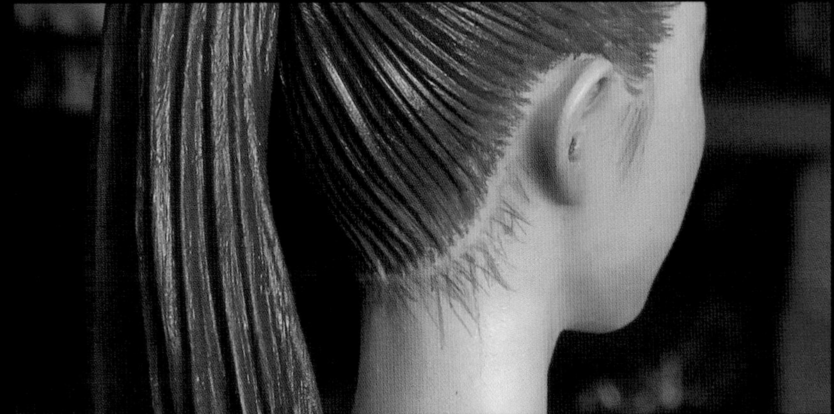

2 有很多男性（以及一部分的女性）會對馬尾造型產生怦然心動的感覺。這是因為這個髮型露出了平常被遮蓋而看不見的後頸部以及鬢毛，以致讓人感受到些許情色性感的衝動。就田川 弘的個人觀察，男性的模型製作者對於鬢毛與汗毛的重視程度還算很少，女性的模型作者相對比較重視這個部分。這可能是因為女性在日常生活就可以觀察到包含自己在內的頭髮等部位的機會較多的關係。如果在人物模型身上加入髮際線這類纖細的表現，能夠呈現出更為寫實擬真的感覺。如果身邊沒有可以觀察實物的機會的話，除了在網路上搜尋圖片之外，也可以觀賞具象畫或是超級寫實主義畫風的人物畫，這樣還可以看到具體的描繪筆法技巧，會比觀察真人模特兒更容易了解如何呈現這些部位，算是學習的捷徑。

Many men (and some of the women) feel emotional about ponytails. This is probably because they usually feel a slight eroticism on the hidden nape of the neck and loose hair. There are still few male modelers who express loose hair or downy hair, but there are more female modelers do. This is probably because women, including themselves, have so many opportunities to see them on a daily basis, but by adding this soft expression to their hairline, they can feel more reality. If you can't get a chance to see the real thing, searching for images on the Internet for figurative painting and portrait called super realism. It is easier to understand than to see the real person because it is drawn more concretely.

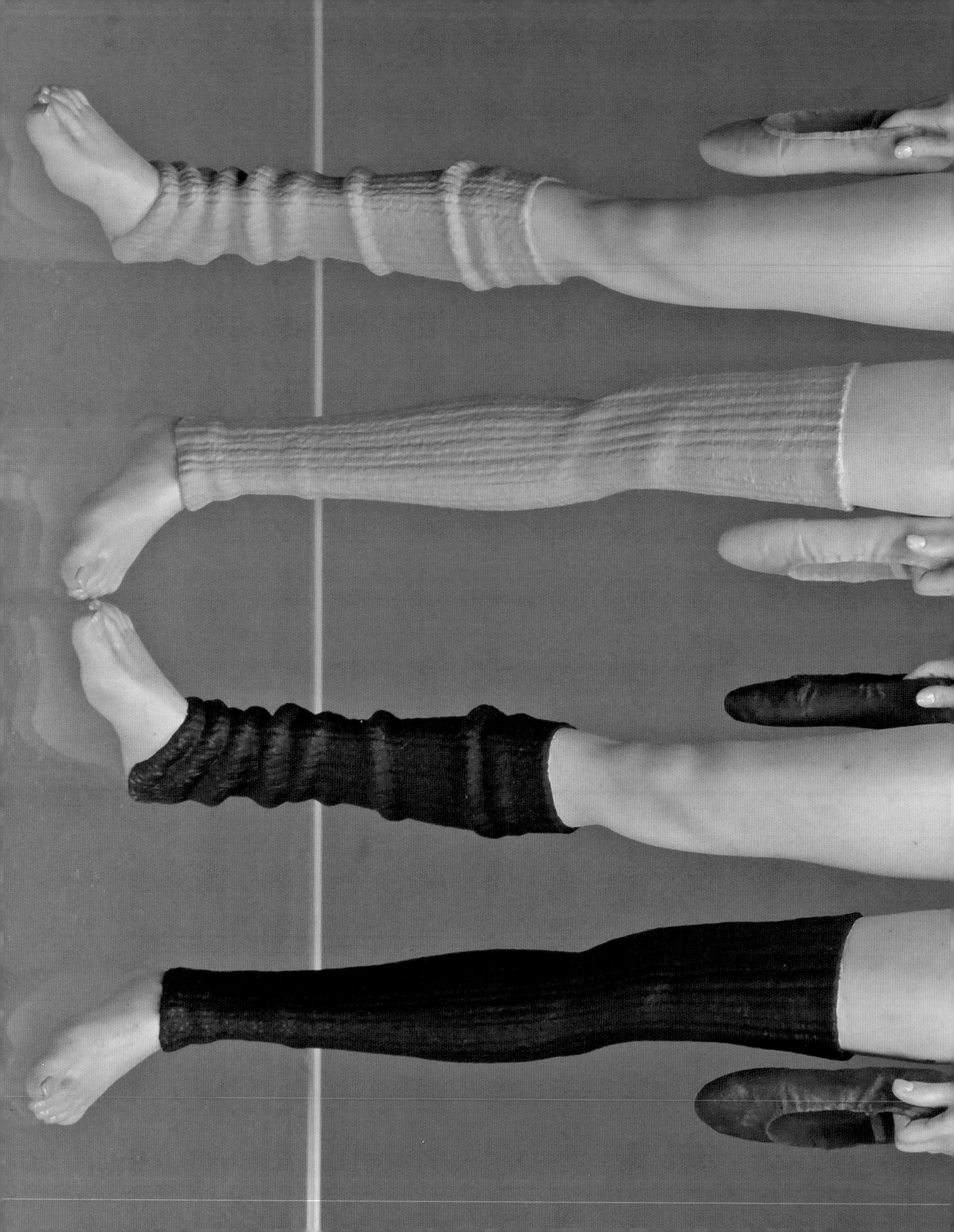

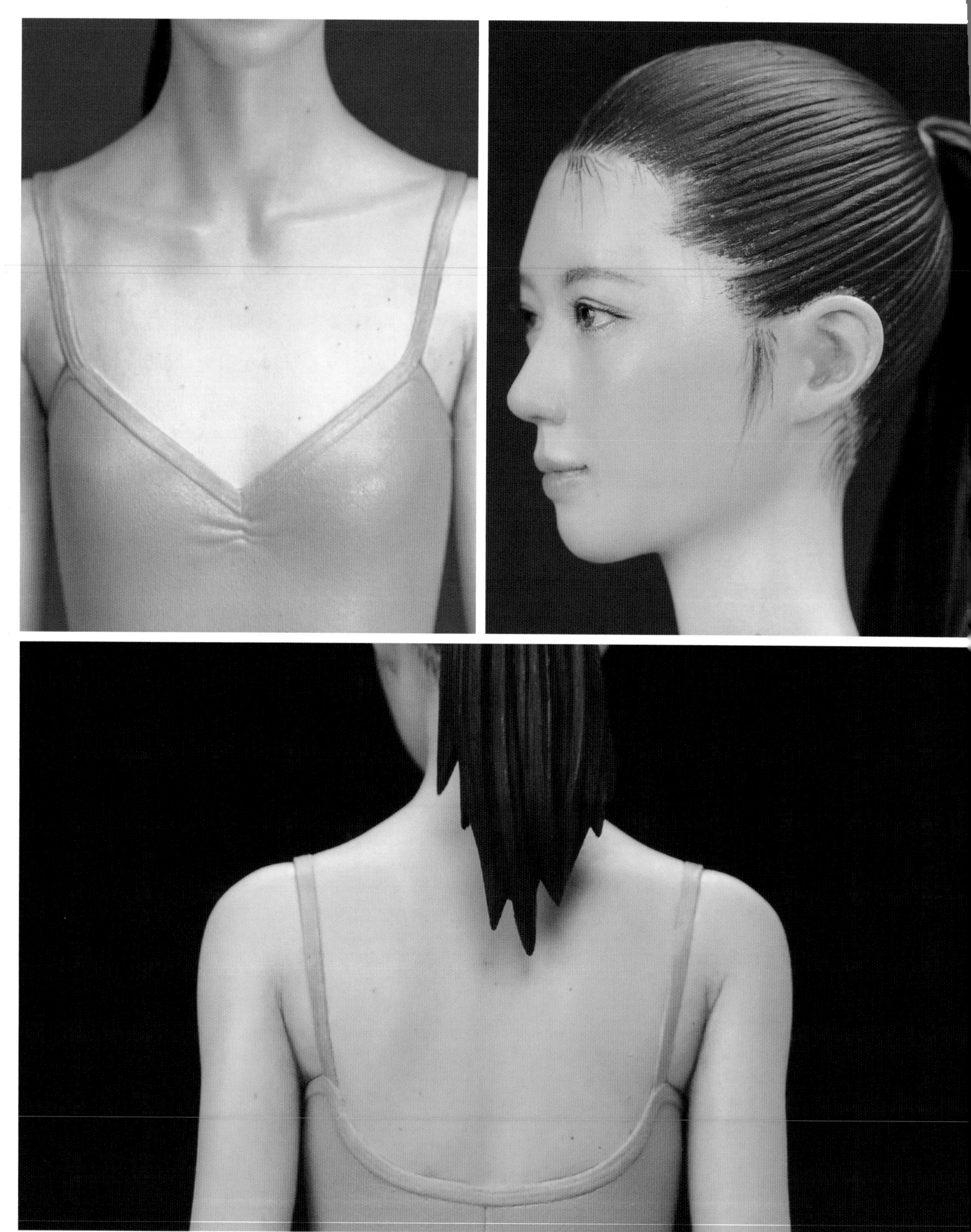

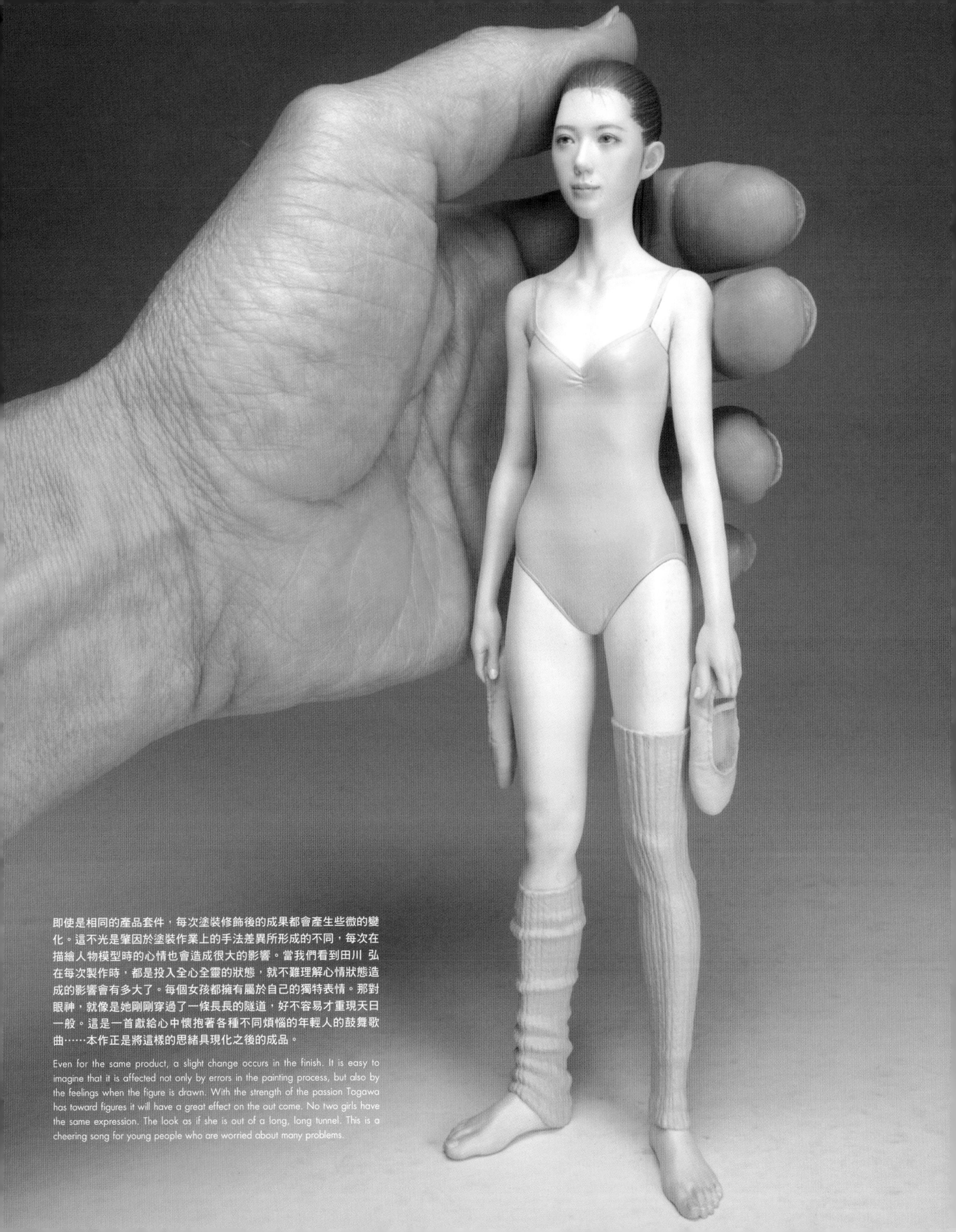

即使是相同的產品套件，每次塗裝修飾後的成果都會產生些微的變化。這不光是肇因於塗裝作業上的手法差異所形成的不同，每次在描繪人物模型時的心情也會造成很大的影響。當我們看到田川 弘在每次製作時，都是投入全心全靈的狀態，就不難理解心情狀態造成的影響會有多大了。每個女孩都擁有屬於自己的獨特表情。那對眼神，就像是她剛剛穿過了一條長長的隧道，好不容易才重現天日一般。這是一首獻給心中懷抱著各種不同煩惱的年輕人的鼓舞歌曲……本作正是將這樣的思緒具現化之後的成品。

Even for the same product, a slight change occurs in the finish. It is easy to imagine that it is affected not only by errors in the painting process, but also by the feelings when the figure is drawn. With the strength of the passion Tagawa has toward figures it will have a great effect on the out come. No two girls have the same expression. The look as if she is out of a long, long tunnel. This is a cheering song for young people who are worried about many problems.

這是田川 弘的作業桌，從基本的組裝到塗裝，都是在這裡進行。姑且不論田川 弘以塗裝為主要工作的塗裝師身份，即使以一般模型製作者諸君為標準，田川 弘的工作環境的特徵仍是塗料以及其他工具的數量「壓倒性的少」。此外，因為油畫顏料都是擠在紙調色盤上使用的關係，並沒有出現在照片上。另一個的理由就是每當完成一道步驟，田川 弘都會先整理收拾之後再進入下一個作業。照明只有用來照亮手邊的兩盞檯燈，並沒有來自正上方的照明。PC 電腦的螢幕和鍵盤是用來電子郵件連絡、上網搜尋、收集資料等等這些一般用途。

Tagawa's work desk handles everything from basic assemb to painting. He may be a finisher, who mainly deals wi painting only, but items on the desk are very few compare to other finishers and modelers in general. Oil paints are n shown in the image because those are used on paper palle And, of course, one of the reasons is that once the process done, those items are put away and he moves to next ste There are only two lights to illuminate his hand and no li from directly above. PC monitors and keyboards are used sending e-mail, searching and data collection.

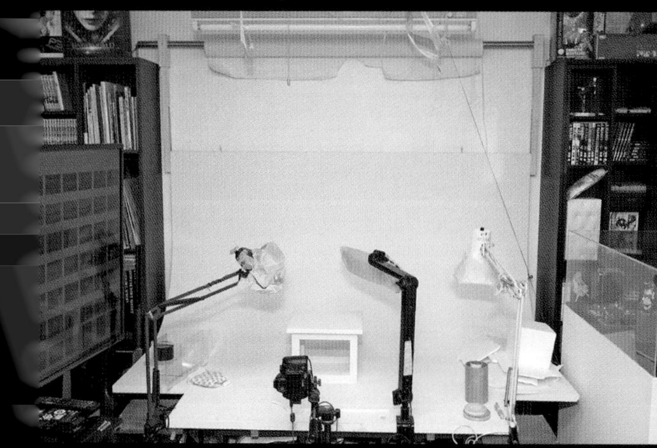

這是常設準備好的攝影棚。背景紙是超大張的白色紙。相機使用的是 Nikon 的 D5000 單眼相機。鏡頭使用的是附屬的 18-55mm 變焦鏡頭。偶爾也會使用 52mm 的特寫鏡頭拍攝 1～2 張特寫照片。

A permanent photo booth. The background paper is white and quite large. The camera is a Nikon D 5000. The attached lens is 18 -55 mm. 52 mm lens for close-up shots.

使用尺寸大到嚇人的 50 吋電腦螢幕來確認拍好的照片。除了可以拿來確認製作途中的模型外觀之外，也可以用來進行模型完成品的色調補償、顏色對比的調整等等作業。

He checks his photos with this 50 inch monitor. In addition to checking the surface of the model in the middle of production, he use it for color tone correction and contrast adjustment.

裝設在廚房一角的塗裝區。以底漆補土做底層處理以及空氣噴筆的噴塗作業都是在這裡完成。廢氣藉由排氣管連通到換氣扇，排出到屋外。

A painting booth in the kitchen. The surfacer spraying and the airbrush paint are performed here. The ventilation fan pull the air to outside.

冷藏庫中放了建議存放在冷暗處的瞬間接著劑，以及紙調色盤上以保鮮膜包覆保存的油畫顏料等等。冰箱中全都是模型相關的消耗材料。

The refrigerator contains only consumable materials related to models, such as instant adhesives recommended to be stored in a cool dark place and wrapped oil colors on paper pallets.

田川 弘作品誕生的工房

Tagawa's workshop

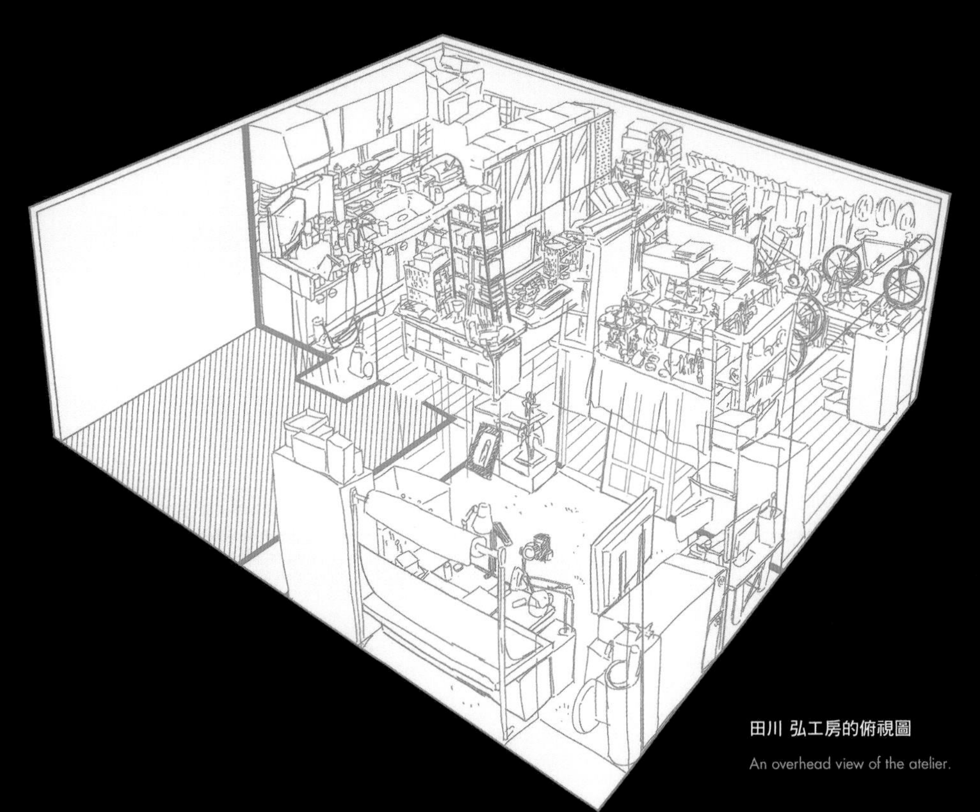

田川 弘工房的俯視圖

An overhead view of the atelier.

放置於室內各處的展示盒中是過去完成的作品，以及將市售人物模型的重新塗裝後的作品。此外還有一部分辦展覽時使用的平面作品。

In the exhibition cases placed at various places, In there are his completed works, repainted figures, and some of the works used in the exhibition.

以前曾經騎過一陣子的重型機車，不過現在是樂在騎單車的生活。因為自己本身就是引擎動力的關係，所以也收到了徹底管理體能狀態的效果。

He used to ride a motorcycle, but now he enjoys riding a bicycle. Since the engine is himself, he has taken thorough care of his health.

收集模型槍是田川 弘眾多興趣的其中一種。由於從以前到現在收集的方向比較集中的關係，東西雖然多，並不會顯得雜亂。

A collection of model guns, one of Tagawa's many hobbies. There are many hobby related things he collected in the room, but it doesn't look cluttered.

趁著自宅改建的時候，將以前居住的空間改造成工房使用。房子裡被工作桌、攝影棚、模型完成品的展示空間，還有與自己的興趣相關的物品塞滿了整個空間。

He uses his former residence as hisstudio when he rebuilt his house. It ispacked with everything from a workbench, a photo booth and exhibition space for finished models.

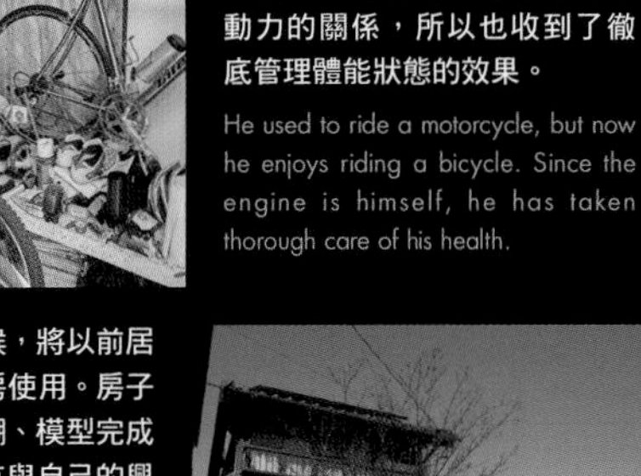

田川 弘趁著自宅改建的機會，將過去生活在其中的獨棟房子直接改造成工房使用。一樓的大部分都規劃作為模型製作所需要的空間。工作桌選用相對精簡不佔空間的款式；塗裝區考量到牆上本來就裝設有換氣扇的關係，決定直接設置在廚房的一隅。此外烘碗機和冰箱也因為尺寸大小的關係，一樣放在廚房裡。雖然分配了一整個房間的空間給攝影棚使用，但這是因為在同一個房間的角落可以用來裝設 PC 電腦，用來讀取相機拍攝作品照片，並顯示在大型的電腦螢幕上進行照片的加工，調整色調及明度，作業起來很方便。每個製作步驟所需要的作業空間都完全獨立，可以增進效率。將一整棟房子整個拿來當作工房使用，相信對有心想要創作的人來說是簡直像夢一般的環境，而那些栩栩如生的女孩們，正是在這樣的環境中誕生來到這個世上。

With the opportunity to rebuild his house, he was able to use his old house as his studio and atelier. Most of the space on the first floor is used for modeling. The workbench is arranged in a relatively compact space, and the paint booth is originally a part of the kitchen because it is equipped with a ventilation fan, and a dryer and a refrigerator are installed in the kitchen due to the size. One room is used just for the photo shoot exclusively, which is convenient for taking photos in one corner of the room and loads them onto a PC in the same room for processing the images and adjusting color and brightness on a large monitor. Each booth is completely independent and efficient. These lively girls are born out of an environment that is almost like a dream for hobbyists.

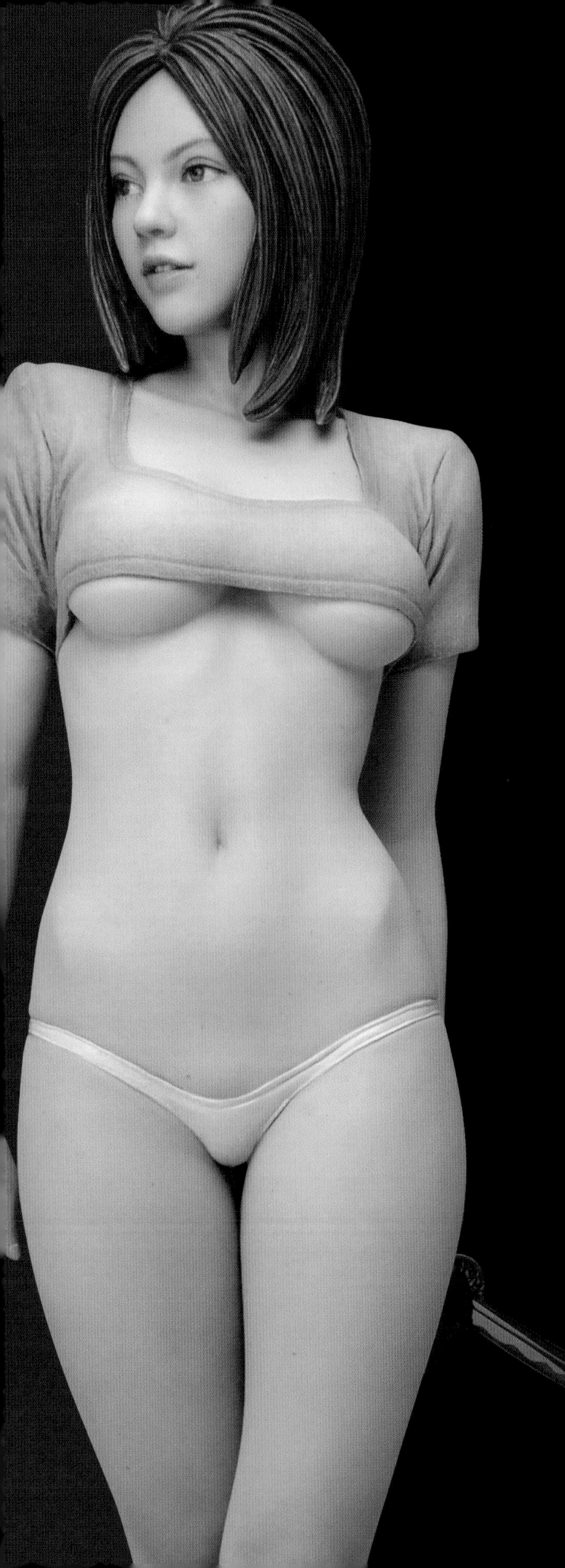

WJ-612

2016 年 1/6 比例 atelier iT 樹脂鑄型套件 原型 林 浩己

2016 1/6 Scale atelier iT Resin cast Kit Sculpted by Hiroki Hayashi

由於一直無法找出人物基於什麼理由擺出半裸＋左腳稍微向前的姿勢，這件作品從暫時組裝到完成為止，足足花費了六年的時間。在本作之前，曾經製作過讓她拿著自動手槍化身為女殺手的嘗試；這件算是第 2 彈作品，改讓她拿著日本刀，接著便一口氣將整個故事背景建立起來了。她是一個女殺手，利用自己的美色接近目標「一位地下世界的重要實業家，並在一夜春宵過後，正打算安靜地出手襲擊對方……」。T 恤是以 Magic-Sculpt 製作，女短內褲是以刮削 GK 白模的方式改造，同時參考了女性角色扮演者牛島良肉（うしじまいい肉）所企劃生產的服裝款式製作而成。

He could not find the reason for this half-naked posing so it took six years from the temporary assembly to the completion. Before this work, he made a female assassin with an handgun, and also one with Japanese sword to complete a back ground story. Assassin wants to quietly attack the dark tycoon after spending a night together.The shirt is made with magic sculpt, and the shorts are shaped by scraping, both of which are made with reference to the costumes produced by Niku, a female cosplayer.

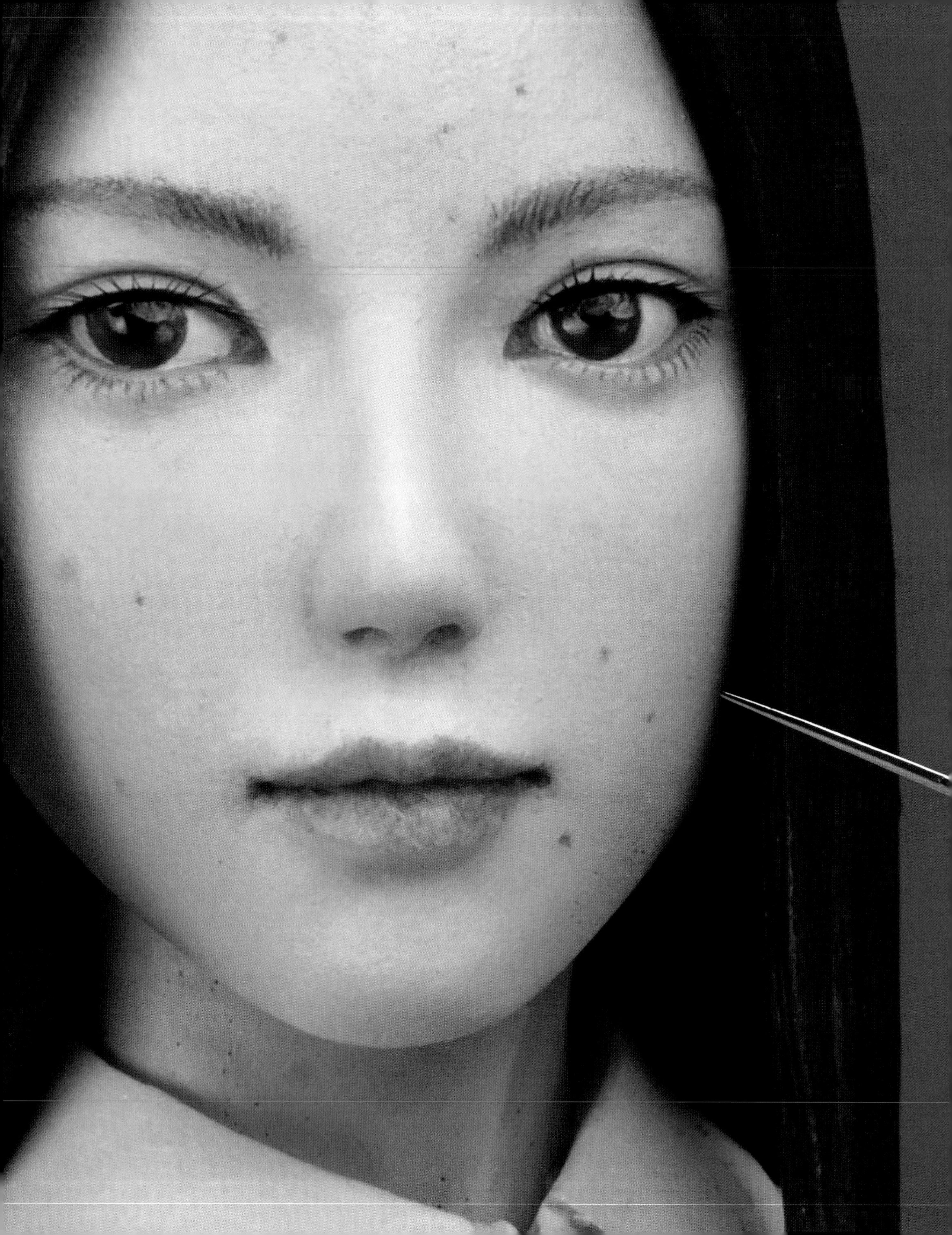

釵 JK

2017 年 1/8 比例 atelier iT 樹脂鑄型套件 原型 林 浩己

2017 1/8 Scale atelier iT Resin cast Kit Sculpted by Hiroki Hayashi

為什麼一位女高中生手拿的不是書包或是網球拍，而是拿著釵這種武器擺出帥氣的姿勢？再加上身邊還跟著一隻柴犬幼犬（套件中的附屬品）？田川 弘對於這組套件的第一印象就是「這到底是什麼狀況？」。懷抱著滿腦子的困惑，總之先開始製作再說。當田川 弘將裙子大幅剪短後，心中浮現了一個白天是女學生，到了晚上就成了懲奸除惡的 Super JK（超級女高中生）的形象。於是便一口氣進入感情帶入的狀態。除了將原始套件的裙子改短之外，就只有加上睫毛的植毛。裙子上的格紋是先在 PC 電腦上設計圖檔，然後製作成自製的水貼，再黏貼到裙子上。

Why are high school girls posing with SAI in their hands instead of bags or tennis rackets? And a Shiba puppy? What is this? He said that was his first impression of the kit. Although he was confused, he started to build it, and when he cut the length of his skirt drastically, he came up with the idea of a girl with two faces. Student during the day and super heroine that fight against evil at night. The pattern on her skirt was made from data he created with PC and printed on the blank decal sheet.

POLICE
DMV 137
警察
995

禁止進

Spinner 懸浮車女孩

2019 年 1/6 比例 樹脂鑄型套件
原型 猿分室假設所工作室 萩井俊士

2019 1/6 Scale Resin cast Kit Sculpted by Shunji Hagii

這個套件是將電影《銀翼殺手》中登場的警用懸浮車擬人化作品，販售於 2011 年。田川 弘自稱這也是沒能夠完全帶入情感便硬把它完成的作品之一。時間經過了 7 年之後，突然想到這孩子說不定是在角色扮演，將自己打扮成警用懸浮車呢？於是決定要重新塗裝這孩子的肌膚部分，也就成為本件作品了。當初是塗裝成洋娃娃風格的表現，重塗之後決定讓她更接近真人的質感。珐瑯質感的服裝是以 2 液型的超光澤透明漆來呈現，成功得到充分符合角色人物性格的質感。

The personification of the police spinner in the movie "Blade Runner", that is what scalptror Said. When he heard that, he was dubious about it, and he completed it without any empathy. And seven years later, He came up with idea of "May be this girl is cosplaying as a police spinner!? " and repainted her skin for this work. At first, it was expressed as a doll, but in repainting, it is closer to a human being. The enamel texture of the costume is made from two-component urethane clear, and the texture is perfect for the character.

惡魔娘 篠崎

2018 年 1/6 比例 樹脂鑄型套件
原型 ke（小抹香）

2018 1/6 Scale Resin cast Kit Sculpted by ke(comaccow)

深邃的五官輪廓、加上一雙靈活的杏眼，明明看起來就像是身邊可見的美麗佳人，卻在太陽穴的位置生出強而有力的骨角，更不用說背上還長著一對翅膀了。透過友人的介紹，第一次見到這個套件時，就被其壓倒性的造形力所折服，同時也深深受到吸引，非常想要為她塗裝，將生命注入她的體內。人類的肌膚、甲冑、護臂以及異形的骨角及翅翼，同時要求具備多種質感要素的表現，在在刺激著身為塗裝師的內心，激發出創作的欲望。田川 弘在執筆本書的時候，正要投入已經是第 4 件的惡魔娘篠崎製作工程，可見得對其的偏愛有多深。本件是其中有做睫毛植毛處理的作品。

A finely chiseled and almond eyes. With a face like any other, a powerful horn sticking out of the temple and a pair of wings on the back. When he first saw the kit through a friend, he was captivated by its formative power, and wanted to paint it and breathe life into it. Expressing the human skin, armor, arm covers, and many different elements such as horns and wings is very exciting for the model finisher. He loves this figure so much that currently working on his fourth one. Eyelash was transplanted in this one.

survival:05 singer

2019 年 1/8 比例 樹脂鑄型套件 原型 大畠雅人

2019 1/8 Scale Resin cast Kit Sculpted by Masato Ohata

時間是近未來，人類正遭到真實身份不明的物體獵殺著。而那個物體，不知為何到了晚上就會停止攻擊。一開始身邊有很多同伴，但如今只剩下少數幾個人。當大家都入睡了的寂靜深夜，她孤身來到遠離同伴之處，一個人唱著歌，是一首年少時期與父親一起觀看電影裡的歌。身為團體領袖的她，經常處在必須得表現堅強的緊繃狀態，此時是她少數能夠做回自己的時刻，隨著心情的放鬆，淚水也不禁流滿了臉頰。而田川 弘，也是一邊流著眼淚，一邊完成了這件作品。吉他使用的是 MEDIA FACTORY 公司的 Ovation 圓背吉他完成品。左手腕及手指改造成按著 C 和絃的姿勢。

In the near future. People who are chased by mysterious being. Somehow it doesn't attack at night. At first there were a lot of friends, but now I have only a handful. In the middle of the night, when everyone was asleep, she sang a song from a movie she watched as a child with her father. A moment of relaxation for the leader comes with tears running down her cheeks. Tagawa himself finished this work with tears in his eyes. The Ovation guitar is from Media Factory, and the left wrist and fingers are modified to hold the C cord.

2017 年田川 弘舉辦個展時，將這 4 位由 klondike 製作原型的女孩，佈置在展場的中央位置。這 4 件讓所有觀看者都能夠感受到強烈存在感的造形物，如女神般全身灰白色的 4 人，是個展中極為重要且必須的舞台裝置。幾位女孩恰如其分地發揮了特徵鮮明的指標作用。背景是以幾位女孩的平面作品作為錦上添花，同樣成功地發揮了重要作用。

These 4 figures, scalped by Mr.Klondike were exhibited at the center when Tagawa's solo exhibition were held in 2017. Produced with the idea of placing them in the center of the gallery as the main characters from the beginning. The gallery's basic color, off-white, was used as the basis, and their flat work was decorated with flowers in the background, and they played a major role successfully

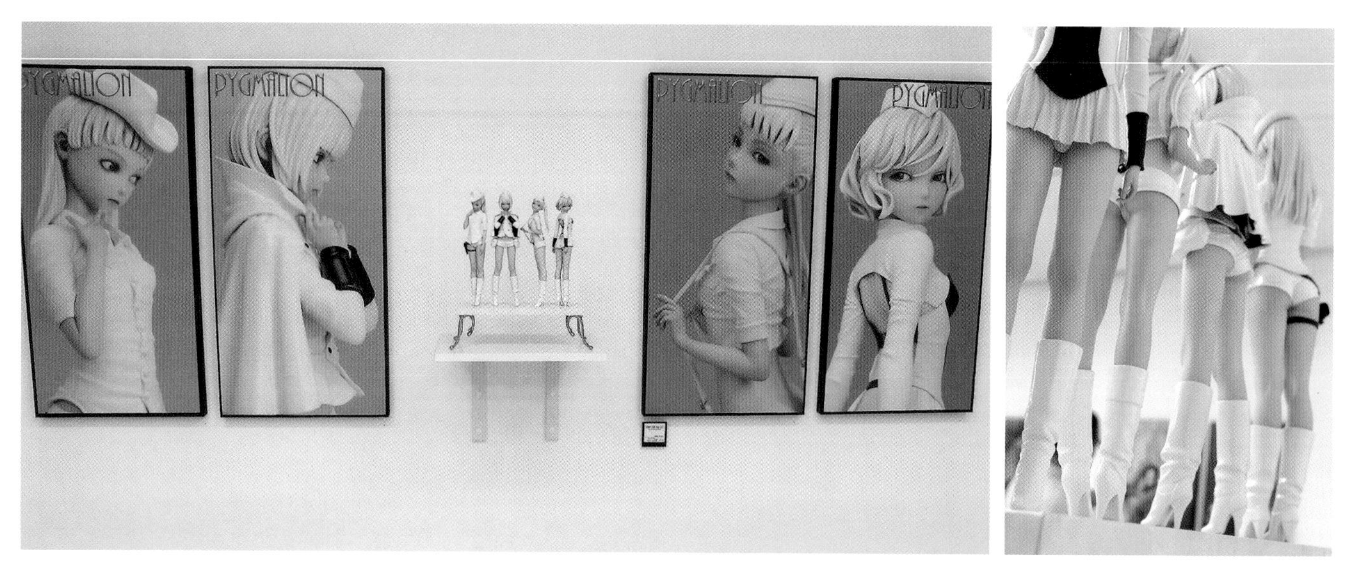

田川 弘所描繪出來的作品本質，藉由這樣的展示方式，綻放著異彩。這是長期參與展示佈置以及舞台佈景的「空間造形」工作，所培養出來的空間設計美感。來場參觀者拍攝的作品及展場的照片放上 SNS 社群媒體分享的一瞬間，等於又誕生了新的平面作品。而這正是田川 弘想要達到的目標效果。

The essence of Tagawa's works blooms in this exhibition. His sense of space design was developed by working on "spatial modeling" such as displays and stage sets. Concentration and separation of the display. Light and shadow by illumination, The fusion of 3 dimentional figures and flat space. This space is the Tagawa's world.

EXHIBITION METHOD

對於展示方法的講究

Exhibition Method

這裡是位於名古屋．伏見的「Gallery 名芳洞」。最初在這裡舉辦的人物模型作品個展是在 2013 年。而 2017 年的這場展覽，是田川 弘第 3 次的個展。展場的主色彩是灰白色。精心的佈置，襯托與提升了每件展品的完成度。試圖要讓來場參觀者都能感受到整個展場空間本身也是一件「作品」。在展場的深處，佇立著這 4 位女孩。平常總是先思考故事背景，然後才開始製作的田川 弘，賦予這 4 位女孩的，只有「這個空間的主宰者」這個單純簡潔的設定。會場中的燈光照亮著地板，如同參拜通道一樣，引領著參觀者步入位於展場最深處的 4 人所在神域。當參觀者站立於 4 人面前時，整個空間呈現出來的氣氛，不禁讓人思考此處她們所代表的涵義為何。4 位充滿靈性的人形（人物模型），彷彿能夠帶給每一位參觀者某種心領神會，而這正是田川 弘所希望呈現的。

"MEIHOUDO GALLERY" in Fushimi, Nagoya. His first personal exhibition of figures was in 2013. His 3rd solo exhibition was held in 2017. The theme color is off-white, making each piece more complete. The idea was to present visitors the gallery space itself as a "Works". There were four figures in the back of the hall. Tagawa usually creates the story for his works, but the only purpose for this 4 girls were to control this space. At the back of the hall, the lighting illuminates the floor like the approach leading to the four girls in the sacred area. Each viewer stands in front of this four girls and they are in a space that makes them think why those 4 girls are there. The idea is to receive something from these soulful dolls (Figure)

Gallery 名芳洞

名古屋市中區錦 1-20-12 伏見ビル B1F

Konny

Erde

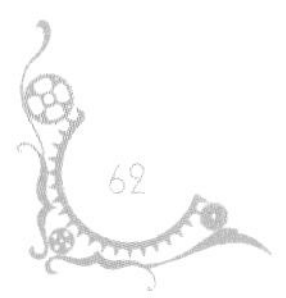

Luce

Soil

2017 年 1/6 比例 樹脂鑄型套件 原型 Klondike

2017 1/6 Scale Resin cast Kit Sculpted by Klondike

WJ-606

2016 年 1/6 比例 atelier iT 樹脂鑄型套件 原型 林 浩己

2016 1/6 Scale atelier iT Resin cast Kit Sculpted by Hiroki Hayashi

本作是在 2002 年曾經製作過，並於 2016 年全新塗裝的作品。當年的田川 弘每天都在自問自答，猶豫著是否要辭掉目前的工作，以人物模型塗裝師的身份重新再出發。而在 SNS 社群網站上認識的朋友們，以及人物模型業界相關的人士，都很願意認真地幫助田川 弘找出這個煩惱的解答。這些朋友每一句發自內心的意見，毫無疑問的，最終成為支持著田川 弘的原動力，並且造就了現在的塗裝師田川 弘。這件作品完成後，田川 弘將其放在拍賣網站上競標，藉此測試自己作品的市場價值，對於田川 弘來說是非常有指標意義的一件作品。

This work was completed once in 2002, and was totally repainted in 2016. At the time,Tagawa asked himself every day whether he would be able to quit his previous job and work as a figure finisher. His friends on SNS and figure industry were the ones who really answered his worries.There is no doubt that every word they said was the driving force behind Tagawa as the finisher. This work was auctioned and measured his value. and became a memorable piece for him.

CAPTAIN PARADISE
凱特琳少尉

2015 年 1/6 比例 樹脂鑄型套件
原型 田中快房

2015 1/6 Scale Resin cast Kit Sculpted by Yoshifusa Tanaka

當田川 弘對著鏡子做出和這件作品相同的表情時，他在鏡子裡看見了自己臉上浮現的皺紋。雖然心裡想著美少女人物模型臉上似乎不適合出現皺紋，但因為沒有人曾經這麼做的關係，反而讓他想要試著去挑戰看看這樣的表現手法。事實證明皺紋並沒有減損人物的美麗，反而更加襯托出她的美貌。喜怒哀樂，所有的表情都會伴隨著皺紋產生。只要能夠好好地駕馭乍見之下屬於負面要素的皺紋表現，就能成功地呈現出人物模型的豐富情感。製作上的改造包括了加上睫毛植毛、將股間的凸起形狀削平、帽章使用飾刻片裝飾、身體的光澤使用 2 液型 Urethane clear 聚氨酯透明高光漆來呈現。

When he looked in the mirror with the same expression as this figure, he saw his wrinkled face there. He thought that wrinkles don't match with beautiful girls figures, but wanted to express something that no one else did, so he challenged himself. The wrinkles made her more beautiful. Wrinkles appear in every expression of emotions. Only by overcoming this seemingly negative factor can a figure's emotional expression succeed. In the production, eyelashes were implanted, the groin protrusion was removed, the cap badge is Photo-etched parts, and the body gloss was expressed by 2 component urethane clear.

Melty Snow
—白雪公主與七宗罪—

2014 年 無比例 樹脂鑄型套件
原型 石長櫻子（植物少女園）

2014 nonScale Resin cast Kit Sculpted by Sakurako Iwanaga

田川 弘還是藝校學生的時候，就曾經深受古董娃娃及創作人偶的魅力吸引。雖然有著人類的形體，但並非人類，而是無機質的存在。然而像這樣連失去靈魂的空殼也不是的事物，是否被允許存在於這個世上呢？此套件的原型師石長櫻子的作品，充滿了像這樣的濃厚的頹廢氣氛。生鏽的鐵籠加上腳鐐、病態般的蒼白肌膚、鮮紅的嘴唇以及白雪公主彷彿看透一切的眼神。田川 弘使出渾身解數的本領，在無機質的物體中注入靈魂。可說是呈現出了真正價值的作品。

He has been attracted to antique dolls and creative dolls since he was an art student. An inorganic being in the form of a human being but not a human being. Is it allowed to exist in this world? Juts like this figure, sculptor Ishinaga's works are full of this decadent atmosphere with rusty cages and shackles, pale white skin and red lips. Her eyes see through everything. This is a work that can be said to be the true peak of Tagawa's ability to give soul to inorganic things.

香草色的早晨

2016 年 1/6 比例 樹脂鑄型套件
原型 田中快房

2016 1/6 Scale Resin cast Kit Sculpted by Yoshifusa Tanaka

場景是在保留著老舊街景的狹窄巷道內「法國。星期天」較晚的上午時間。柔和的日光。「唉唉……頭好痛。昨天喝的酒一點都沒退……」尖銳、高亢的孩童嬉鬧聲。鄰居每天都要上演一遍的夫婦吵架。完全不懂得看場合的鴿子鳴叫聲。走出來到位於 5 樓的陽台外，遠遠的天空可以看到愈飛愈小的客機。沒有預定行程的假日。「啊啊，肚子餓了」。在開始組裝之前，田川 弘的腦海裡不斷地浮現這樣的故事情節。裝扮重點的項鍊使用的是 AFV 模型用的蝕刻片。

A narrow alley with old townscape. Late morning Sunday in France. The soft sunlight. "Ah ...... I have a headache. I can't get rid of yesterday's liquors at all ......。" High-pitched children's voice. Neighbors quarrel everyday. the carefree chirp of a pigeon. Go out to the balcony on the fifth floor, you can see a passenger plane flying through the sky. A holiday without an appointment. "Oh, I'm hungry." He said that this kind of story began to spring up even before the assembly.The necklace uses photo etched parts for AFV models.

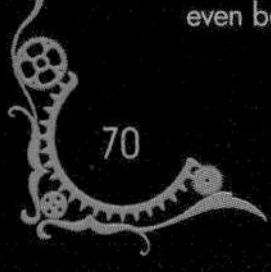

キツネツキ狐憑依

2019 年 無比例 樹脂鑄型套件
原型 イノリサマ

2019 nonScale Resin cast Kit Sculpted by Inorisama

這個套件到目前為止已經製作過 3 件。開始製作前一定要先思考故事情景的田川 弘，對於這個套件一直沒有情節設定的靈感，苦惱了很久。最後好不容易得出的答案就是 Cosplay 角色扮演。這是一位打扮成被狐仙附身少女的角色扮演者。與被狐仙附身這個某種程度算是禁忌的題材兩相對照之下，肌膚的紅潤、細小尖利的犬齒，在在表現出她的幼小年紀以及捕捉獵物進食的生命力。簡潔的白色服裝以及頸部裝飾的紅色，讓人聯想到稻荷神＝狐狸的形象。

He made this same kit three time. Tagawa always thought about the story when he make a figure, but he had a hard time figuring out the situation with this one.The answer is cosplay. A cosplayer dressed as a girl possessed by a fox. Thin, small pointed canine teeth, are an expression of her youth and vitality.The white of the simple costume and the vermilion decorating the neck evoke the image of a fox.

見習魔女

2013 年 1/8 比例 樹脂鑄型套件
原型 藤本圭紀

2013 1/8 Scale Resin cast Kit Sculpted by Yoshiki Fuzimoto

這是一位總是失敗犯錯，每次都被老師責罵的魔法學校學生。老是給周遭添麻煩的小魔女，今天是她畢業前的考試日子。如果在這裡失敗的話，又得回頭去重過一次那個餐廳飯菜難吃的校園生活。令人緊張的一瞬間……結果如何呢 !? 以上就是田川 弘在製作前設定的故事。原型製作者藤本圭紀的完成品參考範例中，衣服的圖樣是描繪成法國象徵主義畫家莫羅（Moreau）的風格。也看到同時期在德國的一位模型製作者將套件製作成法國畫家塞尚（Cezanne）的風格，於是田川 弘就決定在作品壓上金箔裝飾，製作成奧地利象徵主義畫家克林姆（Klimt）的風格。

A magician's school student who always fails and gets scolded by his teacher. Today, she's going through her graduation exam. If she fails here, she'll go back to that miserable school lunch life. her nervous for a moment ......the result is !? Produced with the story as always. The completed figure by its sculptor Mr.Fujimoto, had the costume painted with style of Moreau. At the same time, he saw a German modeler painted it with Cézanne style, and then Tagawa made it with his favorite Klimt style with gold leaf.

Winanna The Hunter

2015 年 1/6 比例 樹脂鑄型套件
原型 タナベシン

2015 1/6 Scale Resin cast Kit Sculpted by Shin Tanabe

本作的原型師タナベシン（Tanabe Shin）的造形，相較於林 浩己所創作的女性來說，大約是年長 5 歲左右的成熟女性形象。在田川 弘的眼中，與其說是可愛，不如說更偏向艷麗的美女風格。不光是表情，骨骼也都已經發育完全，給人正是「成年人（兼美女）」的印象。一系列以精確的素描功力為基底的人物造型帶有獨特的風格，擁有很多的海外粉絲支持者。田川 弘回憶起在製作的時候，因為套件的比例尺寸較大，塗裝的作業也相對較為辛苦。衣服裝飾的皮革、金屬以及布料的質感，是以大多數人這些材質的普遍質感認知來呈現。追加了項鍊來作為吸睛的重點裝飾。閃閃發光，引人注目。

Tagawa says that the sculptor of this figure, Mr. Tanabe's female figures are five years older than the figures created by Mr. Hiroki Hayashi and that they are more beautiful women than cute ones. Not only the facial expression but also the frame is firm, and it gives the impression that "Beautiful Adult". His sculptures are characterized by his drawing power and have many fans from overseas. It was difficult to paint because of its size. The shining necklace is eye-catching on costume with common texture expression such as leather, metal and cloth.

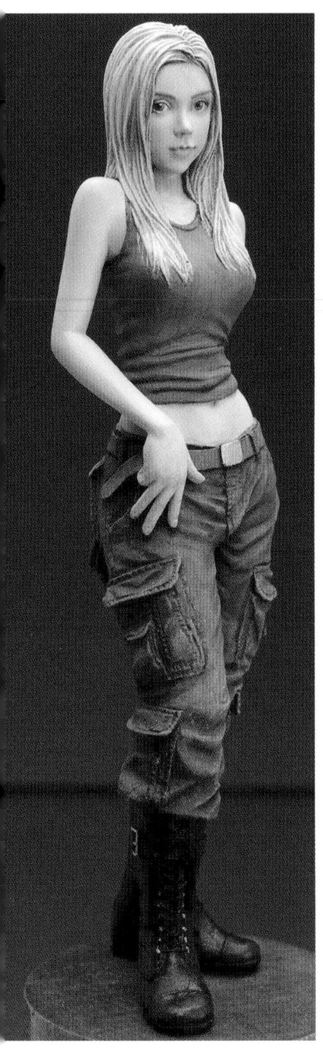

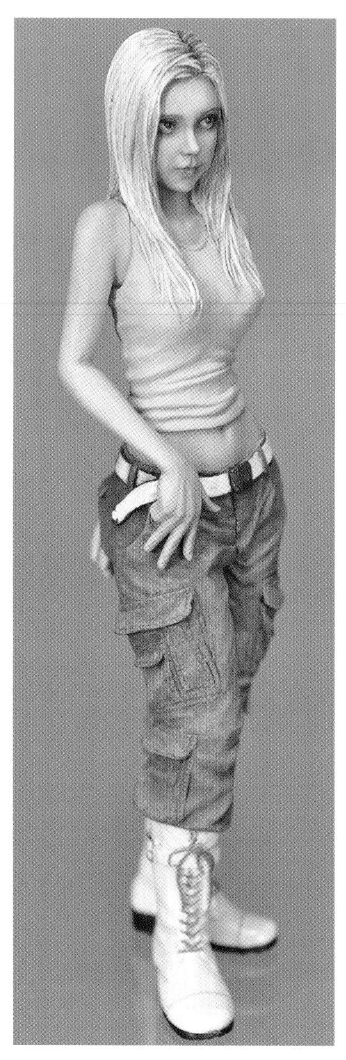

L o p e z

Lopez 貴子，這是以同名同姓的女性為模特兒的人物模型套件（當事者現己改名）。現實生活中她的職業是模特兒以及賽車女郎。長相甜美，屬於蘿莉塔風格的大家閨秀。知道她確定會在 Ma.k.的世界登場時，著實讓田川 弘略感到震驚。

This kit, Lopez Takako, has a real model with the same name. She is a beautiful and cute Lolita type girl who makes a living as a model and race queen. It was a bit of a shock when she appeared in the world of Maschinen Krieger.

Lopez 貴子 From Maschinen Krieger Profile 1 2010~2018 年

1/20 比例 BRICK WORKS 樹脂鑄型套件 原型 林 浩己

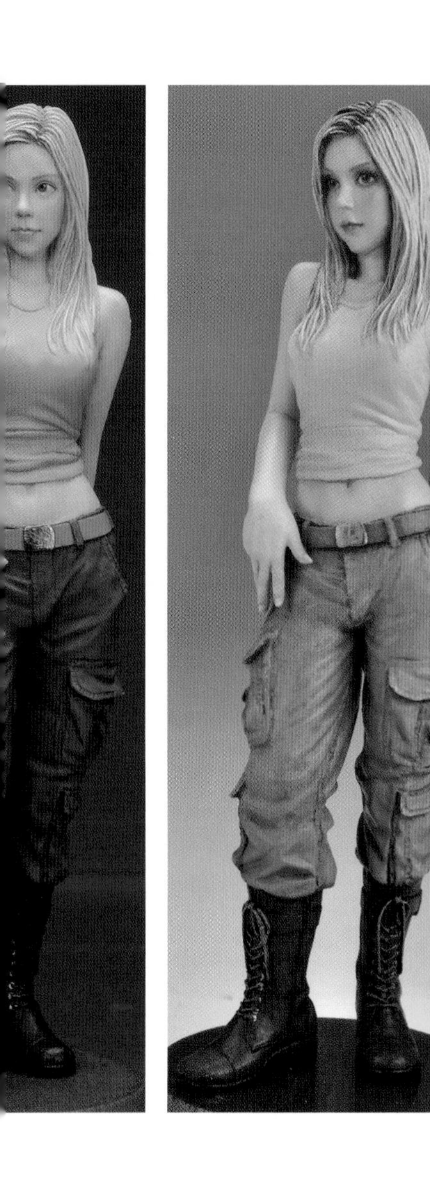

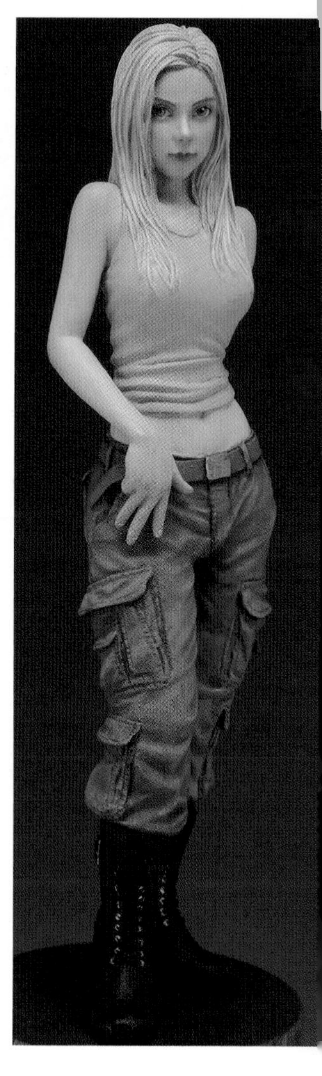

T a k a k o

招牌姿勢很有氣質，而且非常的性感。「以擬真類女性人物模型的觀點，這可說是全世界第一個呈現出「可愛」風格的套件」讓田川 弘讚不絕口。到目前為止一共製作過 9 件作品，而且現在還保有 4 件庫存，說這是讓田川 弘永不厭惓的傑作套件也不為過。

The pose is elegant and somewhat erotic. Tagawa praised highly of this kit by stating "This is the world's first "cute" kit for real women figures." So far, nine have been completed and there are currently four in stock, making it one of the masterpiece kits that you won't get tired of.

2010~2018 1/20Scale BRICK WORKS Resin cast Kit
Sculpted by Hiroki Hayashi

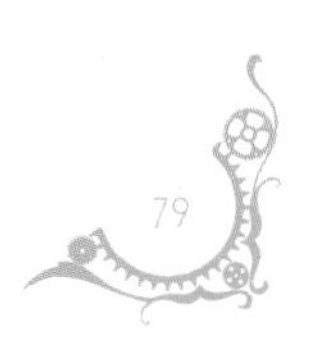

Miracle Girl 12 !
～人氣急速上昇的新秀

2020 年 1/12 比例 BRICK WORKS 樹脂鑄型套件 原型 林 浩己

2020 1/12 Scale BRICKWORKS Resin cast Kit Sculpted by Hiroki Hayashi

田川 弘過去一直都認為 1/12 比例的女性人物模型製作得長相最漂亮、最可愛的套件是林 浩己原型的 HQ12-02（P86），但最近卻迷上了 Miracle Girl 12!這個套件。有一位田川 弘的粉絲支持者告訴他：「我愛上這女孩了。好想和她交往。如果可能的話，還想把她娶回家」雖然只是一件人物模型的作品，看起來卻讓人感覺彷彿她是真實存在的真人一般。簡直就像是有靈魂寄宿在人物模型裡似的。這也難怪。因為就連描繪本作的田川 弘自己也透過顯微鏡的鏡頭，在她身上感受到了愛的存在。

Up until now, he thought that the most beautiful and cute face in 1/12 scale women's figure was HQ-02, sculpted by Hiroki Hayashi, but recently he has been attracted by the Miracle Girl 12. The fan of Tagawa told him "I like this girl. I want to go out with her. I would marry her if I could." Even this is a figure work, they see it as a real person with a soul. Tagawa himself feels love through the lens of his microscope.

いじわるリセットちゃん（壞心眼的莉塞特）

2015 年 1/8 比例 樹脂鑄型套件
原型 ModellerT

2015 1/8 Scale Resin cast Kit Sculpted by Modeller-T

仔細教導田川 弘關於情色表現基本知識的不是別人，正是這個套件的原型師 Modeller T。透過在他手上創作出來的作品，可以學習到世界上的男性都是以怎麼樣的情色眼光？聚焦在何處？而且那些部分都變形轉化成什麼樣態的知識。不可一世的姿勢、扭曲的嘴角以及上揚的眉毛。即使在動畫風格的造形設計中，也能夠感受到內含了紮實的素描功力，呈現出寫實擬真感覺的造型。是一件透過全身的每一處細節，展現出標題的壞心眼氣氛的作品。

It was this figure's sculptor, Modeler-T, who carefully taught the basics of eroticism to Tagawa. He says that he learned from the shapes he created how men's eyes, including their erotic nature, were directed, and how those parts were deformed. A bold looking pose with a slightly distorted mouth and raised eyebrows. It is a shape that has a sense of realism backed up by solid drawing power within the animation-like feel. A work that expresses its title with the whole body.

塗裝師眼中的田川 弘

Tagawa viewed from a painter

mamoru

充分發揮硝基系塗料特性的專業塗裝師

我覺得……在田川 弘的作品中，可以感受得到正在對觀看者訴說什麼似的感情，甚至是意志。有能力超越塗裝的框架，賦予作品達到「存在」的魅力，就我所知僅有田川 弘一人。除了高超的塗裝技巧之外，作品所呈現出來像這樣的特徵，正是引人入勝之處。

From Tagawa's works, you can feel the will to appeal something to the viewer. That is what I felt. He is the only one I know who can go beyond the painting and make the figure "exist" Not to mention the painting technique, I am strongly attracted to such a expression.

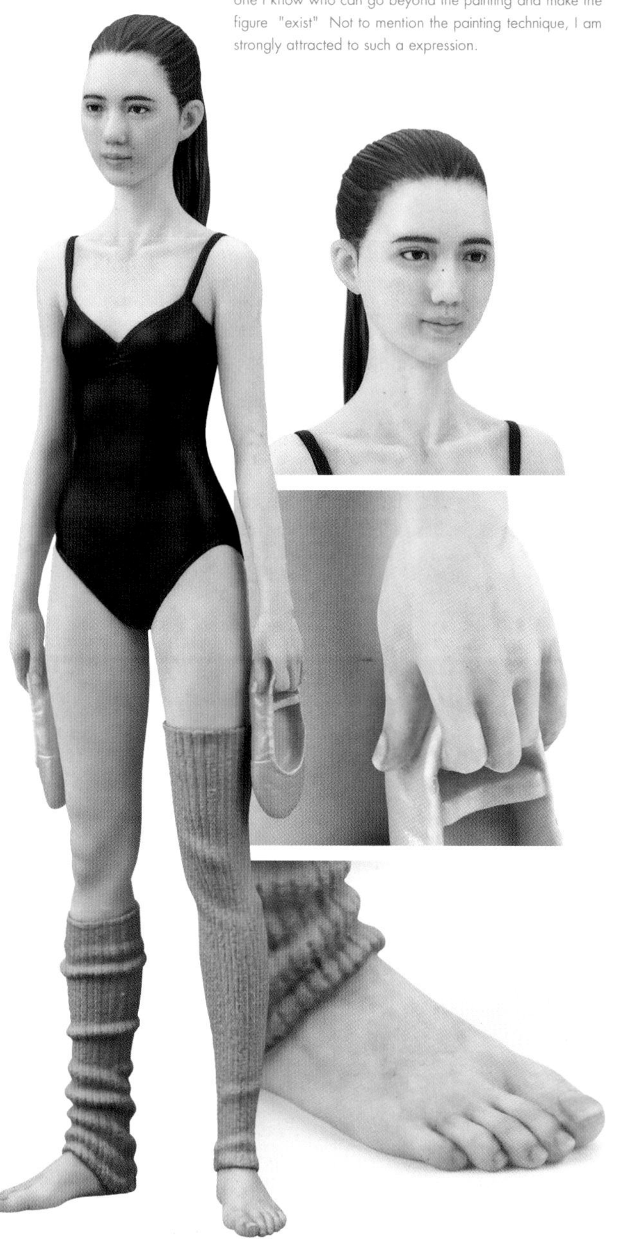

おまちゃん

負責營運 Whichi？公司的新銳女性塗裝師。

田川老師的作品總是教導了我許多事情。藉由將「心」這個眼睛看不到的事物表現出來，能夠讓作品顯得如此閃耀動人。在歡喜、迷惘、甚至是在傷痛當中，田川老師的作品之所以能夠發出耀眼光輝，我覺得一定是作品所呈現出的那股向四周圍訴說著心事的能量。因此我想要在此向田川老師再次說聲「感謝您！」。

Mr. Tagawa teaches us many things with his works. The invisible things are expressed and there's always light behind his woks. I felt joy, hesitation, and sometimes pain. Those feelings radiate all around.So, thank you Mr.Tagawa.

與田川 弘交往過甚的人物模型塗裝師友人是怎麼看待他的呢？
我們邀請了各位塗裝師朋友一同以本書題材的人物模型進行塗裝，並分享各自的感想。

How do the figure painters whom Tagawa is close to see him? We asked each of them to paint the figure that was the main theme of this book and asked what they felt.

村上圭吾

粉絲支持者遍布海內外的水性塗料魔術師

隔空拜入田川老師門下（自稱）已有 5 年時間。我對田川老師的印象只有「完美超人」這四個字。所有技巧幾乎都已經臻於完美。每處細節都給人充滿經驗與智慧結晶的印象。除了本身擁有雄厚的實力之外，再加上對作品投入極為豐富的感情，這樣的態度實在讓人憧憬不已。

I've been pursuing and learning from him for last 5 years. My impression of Mr. Tagawa is that he is a "perfect superman". Perfect technically, and his work gives the impression that his ideas are piled up all over the place. I am longing for his attitude of working with a huge amount of resources and feeling toward his works.

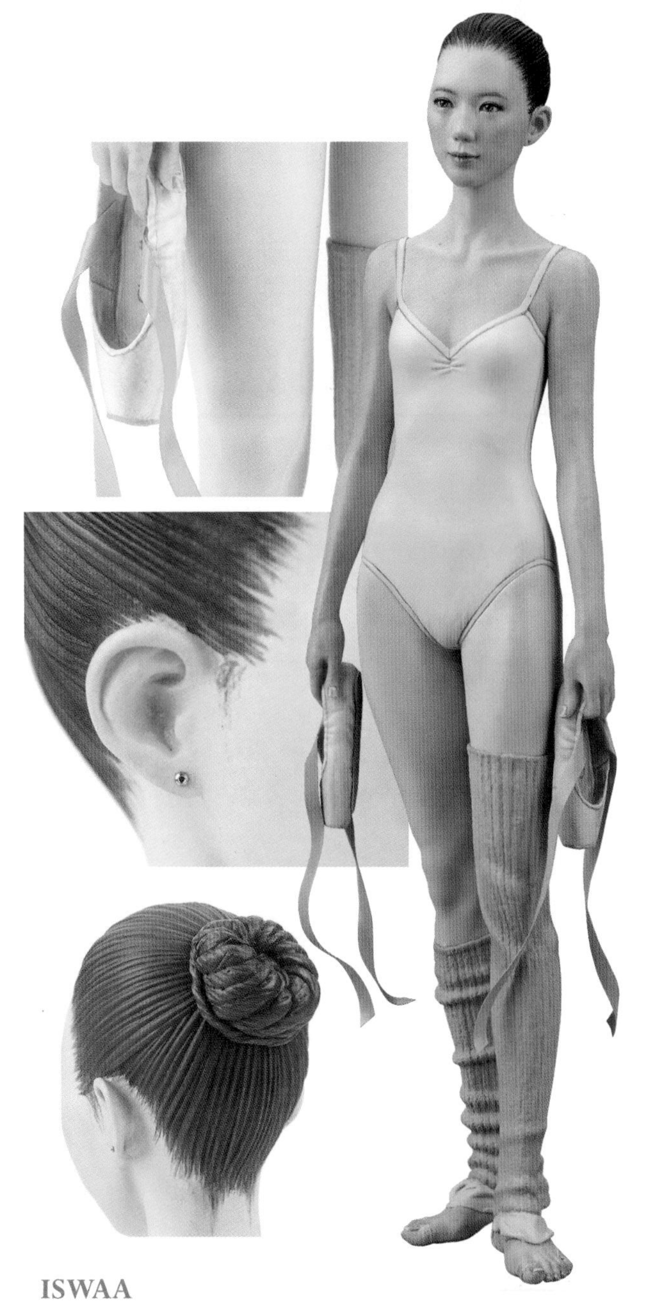

ISWAA

除了塗裝也擅長改造的新銳年輕作家

田川 弘作品的色彩立體深度、精緻的細節描繪以及除了愛情之外再無他想的純粹心意，每一筆畫都在在增添了光輝耀眼的程度。讓人不禁感佩到底要投入多少時間、心力以及思念，才能讓這些女孩誕生問世呢？看了他的作品，沒有人能夠抗拒，沒有人能夠不墜入愛河。

In Tagawa's works, the depth of color and elaborate depiction, as well as the thought of nothing but love, accumulate and shine with each stroke. I wonder how much effort, time and thought are put into making them. There is no reason not to fall in love.

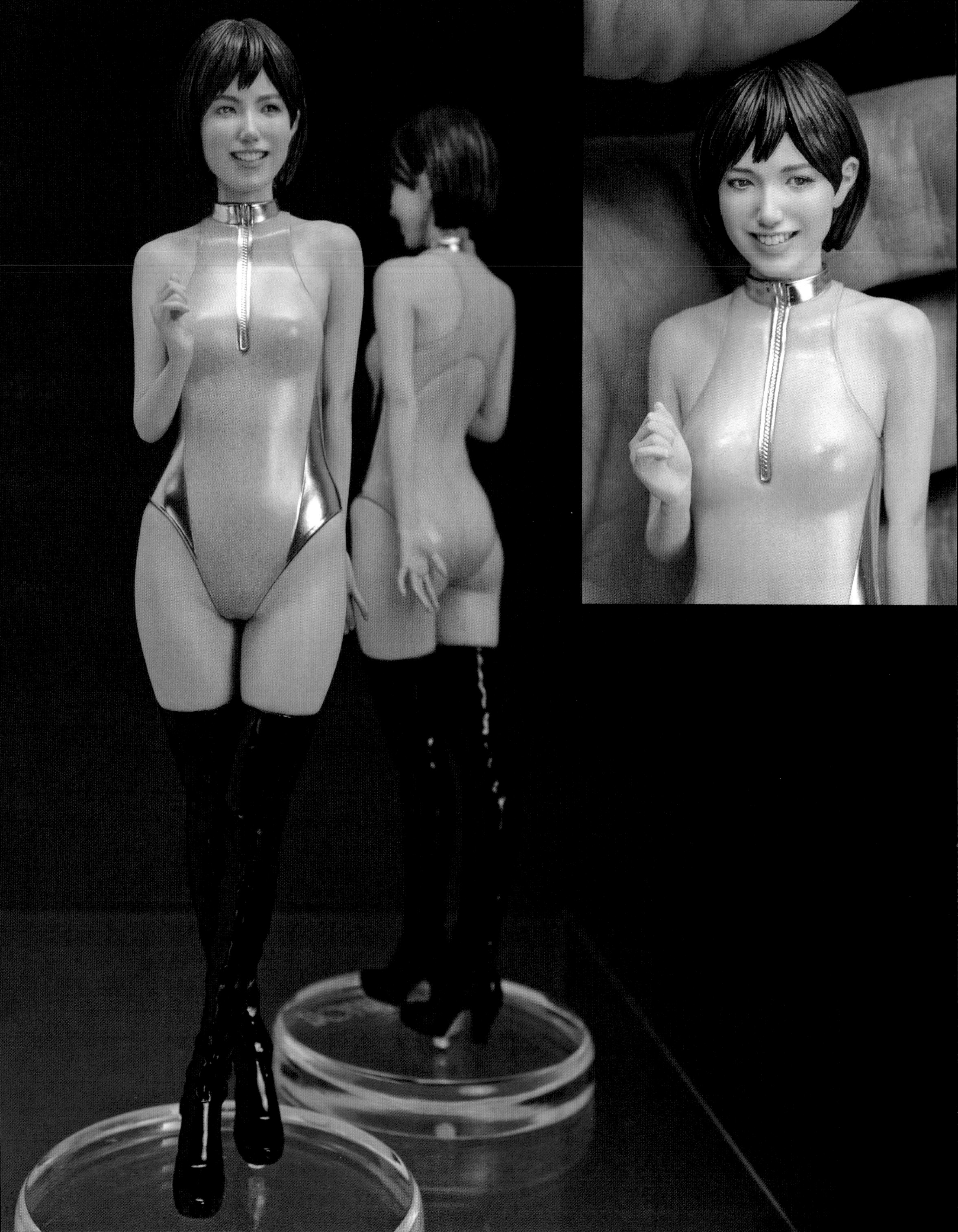

HQ12-02 (P.86-87)

2015 年 1/12 比例 atelier iT 樹脂鑄型套件 原型 林 浩己

2015 1/12 Scale atelier iT Resin cast Kit Sculpted by Hiroki Hayashi

第一次看到這個套件時，覺得左手腕好像些怪怪的，為了要消除這個違和感，便決定將TOMYTEC 公司的 1/12 比例微縮模型槍械「Little Amory」系列的 P226 與戰術手套移植到她身上。滿臉溫柔的笑容背後，手上卻拿著……呈現出如此風情的女殺手完成了。

When he saw this kit, he felt something was wrong with her left wrist. What he came up to get rid of it was the transplantation of TOMYTEC's 1/12 scale miniature firearms 'Little Amory' series P 226 and tactical gloves. With a soft smile on, he completed a lady assassin with a taste.

HQ12-05 (P.88)

2018 年 1/12 比例 atelier iT 樹脂鑄型套件 原型 林 浩己

2018 1/12 Scale atelier iT Resin cast Kit Sculpted by Hiroki Hayashi

看到田川 弘作品的人，經常會提出這樣的問題。「明明只是穿著一般服裝的人物模型，不知道為什麼呈現出滿滿的情色感？」……這我哪裡會知道啊！雖然田川 弘每次都這麼不假辭色的回覆提問者，但其實明眼人都看得出這是作者的心思直接呈現在作品的表情和配色上面了。

People who have seen Tagawa's works often ask him this question. "Why does it feel so sexy even though it's a figure in a normal costume?" "I don't know!" he answers to them, but it is needless to say that the author's heart shows the expression and coloring on the works.

WJ-611 (P.90)

2016 年 1/6 比例 atelier iT 樹脂鑄型套件 原型 林 浩己

2016 1/6 Scale atelier iT Resin cast Kit Sculpted by Hiroki Hayashi

這是將 2007 年製作過的作品，於 2016 年全新塗裝後的作品。一個人獨居的 OL 粉領族，從公司回到家裡，脫掉身上的衣服，一邊自言自語「有沒有什麼好看的電視節目呢？」一邊快速瀏覽著不同電視頻道，如此這般的場景設定。姿勢自然，表情也是田川 弘覺得滿意的作品之一。底褲的圖案是以 PC 電腦自製的水貼。

It was made in 2007 and repainted in 2016. An office lady who lives alone, comes home from work, undresses, says "Are there any interesting programs on TV?" and zaps the channels. This is one of his favorite work with a natural pose and expression. The pattern of the shorts is a decal made with PC.

WJ-60 (P.91)

2016 年 1/6 比例 atelier iT 樹脂鑄型套件 原型 林 浩己

2016 1/6 Scale atelier iT Resin cast Kit Sculpted by Hiroki Hayashi

這是將 2002 年製作的作品，於 2016 年再次全新塗裝的作品。背後的頭髮的塗裝處理是受到本作的原型師林 浩己的製作範例啟發的結果。雖然不想要模仿其他人的技法，但還是很想要學習優秀的技術納為己用。把好的東西偷學過來，經過自己的消化吸收後，再去昇華成為更完美的技法。

It was made in 2002 and repainted in 2016. The back hair paint work was inspired by Hiroki Hayashi's work. He didn't want to imitate anyone, but wanted to own the technique. He steals the good skill, digests it in himself, and sublimates.

林 浩己和田川 弘是在現今最為火熱的擬真類女性人物模型界中，兩位最廣為人知的大師級巨匠。兩人的關係是彼此相互共鳴的舊友，其如同命中注定般的邂逅經過為何呢？

而兩人之間又對彼此存在著怎麼樣的認知？兩位巨匠這場彼此受到對方吸引為必然發展的邂逅，經過多年相知相識後，如今總算有機會在這場對談中說出對於對方的真心話。

——可否請簡單說明兩位是怎麼認識的？

林　你記得嗎？我完全沒印象了（笑）。

田川　我在偶然的機會下在咖啡廳翻閱一本叫做《POPEYE》的雜誌，裡面刊載了林老師的裸體人物模型，當下就覺得「哇……做得好像真的！」而受到感動。因為我很想要親眼看到這些作品，所以就在網路上試著搜尋看看。然後就找到了 atelier iT（林老師經營的個人品牌，以下稱為 iT）的網站。想說「原來製作這些人物模型的是一位叫做林 浩己的人啊」，再看到網站上有他的電子郵件信箱，就姑且一試寫信連絡了。

林　……完全沒印象（笑）。

田川　後來就收到「你好啊～」這種輕鬆語氣的回信，讓我嚇了一跳。好像是個蠻好相處的人呢（笑）。後續這就樣經過 2、3 次的書信往返。

——這大概是多少年前的事情呢？

田川　差不多 25 年前吧。因為雜誌的發刊日期是 1995 年的。

林　我們都聊了些什麼啊？

田川　「我從來沒看過像這樣的人物模型！」之類的吧。雖然當時市面上已經有一些擬真的人物模型出現。但都是像 VOLKS 公司推出的妖鳥死麗濡（在《惡魔人》中登場的角色）這種以動漫角色為造形對象的模型作品。

林　當時女子高中生這類擬真類人物模型真的沒有市場需求。大家都不知道該怎麼塗裝，就算買了也是「該拿她如何是好……」這樣的感覺。在這種狀態下，大型企業根本不會想碰觸這塊市場，最多只有株式會社 REDS（2003 年已結束營業）基於興趣來幫我推出產品。再說當年只要一提到人物模型，一般指的都是動漫角色的模型。即使推出沒有角色故事背景的女性人物模型……也是……哎，你懂的。

田川　林老師在 READS 退出市場後，開始製作了一些可動人偶的頭雕對吧。而且都是一些電玩遊戲的角色人物模型，讓我開始焦慮了起來。我完全沒有涉獵電玩遊戲，一點都不知道林老師在做的角色是什麼……？

——對田川老師來說，是不是希望林老師能繼續製作更多的擬真類人物模型呢？

林　我真正喜歡的是將造形對象製作成擬真的立體塑像。從以前我就以成為一名插畫家為志向，喜歡描繪人物的寫實表情。不過工作上製作的內容基本上都是動漫人物，所以我做得不是太開心。我覺得「如果不做自己喜歡的東西，早晚感性會變得遲鈍！」，於是便開創了 iT 這個自有品牌。過程中確實有一段時期是將精力都投入在製作關節可動人偶的頭雕上，那時對於 iT 的作品就減少了許多（笑）。沒辦法，畢竟製作了也賣不好。是真的都賣不出去……。

——擬真類的人物模型和動漫類的人物模型相較之下，銷售的數量如何呢？

林　雖然我想回答說大約是動漫類的 1/10 左右，但其實可能還不到這個數量。

二位巨匠的幸福邂逅及其真實經過

The joyful encounter between two masters and its truth.

以壓倒性的表現力，在擬真類女性人物模型界倍受肯定，穩居大師級地位的二位巨匠
為各位讀者闡述兩人的相會過程以及迄今的交遊關係

——那本雜誌《POPEYE》關於擬真類人物模型的報導，大約是在 1995 年的 1 月發刊，當時是《新世紀福音戰士（以下簡稱 EVA）》正在流行的時候吧。我記得市場上因此一口氣推出了許多 EVA 相關的人物模型，讓人物模型在市場上就像是一下子取得了一席之地。

林　當時取得一席之地的只有動漫類的人物模型。即使是在 Wonder Festival 的會場上，也只有動漫類的攤位大排長龍，擬真類的人物模型還是乏人問津。在我記憶中，大家都只是路過靠近來瞄上一眼，接著就轉往其他地方去了（笑）。另一方面，可動人偶有很多是和《大英雄》（G.I.Joe）相關的產品，當中有許多人非常努力想要呈現出擬真的效果。聽到他們之間在熱烈討論如何呈現出真實感，我也感到很開心。尤其當我也加入討論後，大夥的氣氛就變得更加熱烈了。也因此促成了我開始製作可動人偶的契機。

兩人的相遇及其過程

——林老師是從什麼時候開始意識到田川老師的存在呢？

林　我每次在 iT 推出新作品後，馬上就會在網站上發表那件產品的塗裝範例。和田川老師開始互有往來應該就是在那個時候吧。當時我在自己的網站上開了一個像是網頁留言版的頁面，兩人便開始在上面交流。

田川　發表在網站上的作品都做得太好了，對我來說非常有參考價值。而且隨著作品發表的件數愈多，製作的品質愈來愈好呢。

林　一開始田川老師針對我的競爭對手意識真的好強烈（笑）。我最早的印象是這人怎麼這麼強勢，真是個討人厭的大叔。主要是因為有一次我在 iT 上發表了 Q 版的人物模型。就是當年《ぴあ》雜誌封面非常流行的那種 Q 版造型。頭部比例畫得很大的那種。我也曾經以「這樣好像也蠻有趣的」的心態製作過類似的作品。沒想到田川老師馬上來連絡說「那樣不對！」。他開門見山就提出「林老師你現在的製作風格是最有特色的，請你堅持下去！」這樣的意見。一開始我還心想「我製作自己喜歡的作品為什麼還得經過你同意啊？」（笑）。

田川　我是真的很希望林老師回來製作擬真類人物模型……。

林　不過，等我冷靜下來一想，才明白到田川老師他寫信來的目的並不是要指責我，而是用他自己的方法來想辦法引導我。所以雖然一開始心裡不痛快，但很快就恢復平常心了。

田川　……我記得當時您還有回信反駁我。

林　是不是像「不許這樣批評我！」之類的？

田川　確實是委婉地傳達了那樣的意思在內啦（笑）。

林　其實到現在也還是一樣的狀態（笑）。田川老師真的是很直率的一個人。自己心裡怎麼想就直接說出口。如果我不能理解他是這樣的個性的話，一定早就分道揚鑣了。想說這傢伙專門來挑釁的……。不過這個人說話是很直率、真誠，而且是真心話，表裡一致，批評別人也不是為了滿足自己的優越感。可以感受到他說這些話真的只是為了我著想。既然如此，冷靜下來後，我也重新思考「果然還是製作擬真類人偶比較適合我吧」。

——這是 25 年前兩位認識後，經過 4、5 年左右的事情嗎？

林　是的。順帶一提，我們兩個真正面對面見到對方，是在 2016 年大阪舉辦的「Modeler's Expo」的會場上。
田川　啊啊，確實那時是我們初次見面。
──咦？兩位認識這麼久的時間，原來一直沒有真正見到對方嗎？
田川　是啊。那段期間也是發生了不少事情。林老師大約在 10 年前，作品的風格開始迷失了方向。那份迷惘只要一看到當時在 iT 推出的產品就能感受得到。啊～林老師該不會陷入迷惘了吧。
林　……是呈現怎麼樣的迷惘狀態呢？
田川　用言語很難形容，有好幾件作品都讓我感覺到「林老師好像失去了活力」。
林　啊啊，你說那個啊……不會吧？你真的有察覺出來哦？（笑）當時除了作業時間被限制得很死之外，還有受到很多其他的束縛……。不過還真虧你有辦法感覺得到。
田川　哎呀……我這麼長時間都在關注你的作品，當然感覺得到囉（笑）。

兩人的合作是從 20 年前開始

──與林老師之間的交流，對於田川老師的作品風格產生了什麼樣的變化呢？
田川　因為我一直很嚮往林老師的作品，所以我的風格基調裡一直都有林老師的影子。不過其實對於我的作品風格產生最大變化的，反而是因為開始接觸到了林老師以外的作品。打從 7、8 年前起，我的作品風格開始有了很大的轉變。當時我結交了許多其他的擬真類人物模型的原型師朋友，開始受到他們的影響。那個時候正是「Fg」（創作者可以將自己的作品照片投稿到網頁，與其他成員彼此交流的社交網站）大為流行的時代。

──林老師有發覺到田川老師的作品風格變化嗎？
林　其實那段時間我們兩個沒有什麼交流。
田川　沒錯。有好長一段時間沒有交流了。就連 Lopez 貴子紅遍全日本的時候，我也不知道原來林老師有製作她的人物模型。
林　我們先前都是透過網站留言版進行交流，不過我因為一些事情，曾經把我的網站關閉過一陣子。自那時起兩人的交流就中斷了。後來經過一段時間，我在尋找製作人物模型的題材時，想到說如果把寫實風格的平面畫作製作成立體人物模型，應該蠻有意思的，接著就想起田川老師早年的畫作中有一件我非常喜歡的作品。於是就主動連絡田川老師洽詢是否可以讓我製作成人物模型來銷售販賣。
田川　正是如此。林老師突然連絡說「我想將你的插畫製作成立體人偶，不知你的意下如何？」嚇了我一跳。「那位林老師居然會主動找我合作？」非常意外的感覺。
林　那件人物模型作品是在 2000 年左右完成的吧。當時在 Wonder Festival 會場也展售了未塗裝的白模 GK 套件。結果完全賣不出去啊（笑）。
田川　對對對（笑）。

林 浩己所觀察到的田川作品之全貌

──最初在《月刊 Armour Modelling》企劃的女性人物模型特集中，曾經向林老師問到「田川老師是位怎麼樣的人呢？」，當時林老師的回覆是「另外一個星球的人。」，能不能請林老師再詳細說明您觀察到的田川老師全貌呢？
林　首先，田川老師他並不是以男性的出發點來做塗裝的。即使是裸體人像的作品，他的出發點也不是來自於男性的欲求。以我來說，當我在塗裝女性人物模型時，我會強迫對方呈現出更接近自己理想中的女性形象。像是「我來讓妳變成一個好女人！」這樣的感覺。不過田川老師則會讓自己成為被塗裝的那一方。「如果是我自己是她的話，我會想要變成這樣的風格！」這兩者是完全不一樣的。田川老師的女粉絲很多，應該就是這種更接近「想要讓自己變得更可愛！」的女性心理形成了共鳴吧。所以在他的作品中看不到男性強勢的自我主張。也因此他和其他男性模型製作者所處的世界不同。我對這點有很深的感受。
──以前我在和田川老師的訪談中聽過田川老師將塗裝的作業用「化妝」這個名詞來稱呼。這裡的「化妝」所包含的語意，似乎也和其他男性模型製作者所認知的不一樣吧？
林　我認為田川老師在這個「化妝」裡面追求的是如同嬰兒或少女般水嫩的肌膚感。所以和我們一般男性所想的「化妝」會有一些不同，應該比較接近女性在追求的目標吧。
──這就是田川老師作品風格的關鍵特徵吧。
林　說的沒錯。特別是在意識的部分。一般來說，人物模型的塗裝有所謂的塗得好或壞，但田川老師的作品，卻無法用好壞優劣來評論。我不認為田川老師的塗裝特別好，再說技法是任何人都可以模仿學習。但他從不放棄努力呈現出人物的可愛之處，總是使出渾身解數一直到最後一刻為止。這就是田川老師令人敬佩之處。一心只想著「我要讓妳變得更可愛」這件事，直到整個製作完成為止。這也就是為什麼田川老師每次完成一件作品都可以很自然，而且毫無顧忌地大聲說「大家快看！我讓她變得這麼可愛囉！」。一般人就算心裡這麼想，也很難說得出口吧？更何況還是個成年男子。（笑）
──如果製作的是女性人物模型，也許就更難說出口了呢。順帶一提，林老師有沒有想要看田川老師製作哪一類的作品呢？
林　應該沒有。我不怎麼去理會外界的事情。我只關心自己要前進的方向。田川老師的塗裝是帶有想要將人物模型製作得更可愛的意識。而我從以前就抱持著「想要呈現出人的真實面貌」這樣的意識。這兩者是完全不同的方向。不過，田川老師的作品風格好像愈來愈朝向人的真實面貌的方向移動，稍微靠近我的領域了。我還想說你別再靠過來了（笑）。
田川　我在塗裝之前都會先設定一個故事情節和人物所處的場景，也許是這樣的思維方式讓我的作品風格逐漸產生了變化吧。
林　其實是我受到那種「先設定故事情節」的方法的影響較多。從幾年前開始，我製作原型的時候都會先思考人物的時空背景，仔細揣摩這孩子心裡在想些什麼，才開始動手製作。結果這份心思也確實傳達到收藏玩家的身上。即使對方感受到的不完全是符合我所設定的場景情節，但總有一部分是我想要傳達的情感。然後收藏玩家會再衍生出屬於他們自己的故事。這麼一來，我們賦予人物模型的就不只是美麗或可愛的外型，連同意志和想法也灌注在其中。像這些我自己從來沒有思考過的事情，都是田川老師教導給我的。以這層意義來說，我能夠和田川 弘這個男人相識，實在是太幸運了。
田川　你別這麼說，太誇張了啦！

（2020 年 3 月，於神奈川縣 濱市）

林 浩己

Hiroki Hayashi

1965 年生，現居於日本神奈川縣。自由商業人物模型原型師。1999 年頃，成立 GK 套件製作公司「atelier iT」。作品風格寫實且優美的造型而廣受到好評。林 浩己原型的粉絲支持者遍佈日本國內及海外，往往在「Wonder Festival」等 GK 模型套件展售會等活動中推出的產品都以極快的速度銷售一空。製作人物模型的信念是「忠於自己所好」。

PYGMALION

畢 馬 龍 P Y G M A L I O N

令人心醉惑溺的女性人物模型塗裝技法

如此美麗，讓人不禁陷入戀情的魅惑女性形象・田川 弘塗裝作品 A to Z

作　　者　田川 弘
翻　　譯　楊哲群
發 行 人　陳偉祥
發　　行　北星圖書事業股份有限公司
地　　址　234 新北市永和區中正路 458 號 B1
電　　話　886-2-29229000
傳　　真　886-2-29229041
網　　址　www.nsbooks.com.tw
E-MAIL　nsbook@nsbooks.com.tw
劃撥帳戶　北星文化事業有限公司
劃撥帳號　50042987
製版印刷　皇甫彩藝印刷股份有限公司
出 版 日　2020 年 10 月
I S B N　978-957-9559-60-7
定　　價　480

如有缺頁或裝訂錯誤，請寄回更換。

國家圖書館出版品預行編目（CIP）資料

PYGMALION 令人心醉惑溺的女性人物模型塗
裝技法 / 田川 弘作 ; 楊哲群翻譯.
– 新北市 : 北星圖書, 2020.10
　面；　公分
ISBN 978-957-9559-60-7（平裝）

1.玩具　2.模型

479.8　　　　109013093

Special thanks

林 浩己
K
藤本圭紀
田中快房
大畠雅人
石長櫻子（植物少女園）
Modeller T
猿分室假設所工作室 萩井俊士
Klondike
イノリサマ
ke
タナベシン
atelier iT
KITA
おまちゃん
mamoru
ISWAA
村上圭吾
吉岡和哉
もやし
凌元臻
Entaniya
BRICK WORKS
榊　征人
牛島好肉（うしじまいい肉）
吉田伊知郎
堀 和貴（ART BOX）

臉書粉絲專頁　LINE 官方帳號